HOUSING AND FINANCE IN DEVELOPING COUNTRIES

Half of the world's urban population lives in poverty and approximately 800 million people occupy substandard housing. This 'housing crisis' has continued unabated despite over twenty years of research and policy. At the forefront of new policy initiatives, confirmed by conferences such as Habitat II in Istanbul, is an effort to give greater priority to finance. The expectation is that the provision of small quantities of finance to low-income households will bring real improvements to the quality and quantity of housing provision.

Housing and Finance in Developing Countries explores the linkages between formal and informal housing finance, drawing upon the lessons of NGO and micro-finance practices. Both public and private formal finance institutions have experienced great difficulty in lending below a middle-income client group, and are often reluctant to lend for the purposes of housing at all. This failure of formal finance to filter down to low-income households, and in particular to women, has led various NGOs and community groups to create and adopt innovative finance programmes, such as informal savings banks and credit rotating schemes. The authors critically assess the impact of these schemes and evaluate links between gender, housing and finance.

Drawing upon experiences in a wide range of developing countries in Latin America, Africa and Asia, *Housing and Finance in Developing Countries* includes contributions from academics and from representatives of NGOs and development organizations, from both developed and developing countries. Providing important insights into how housing provision is affected by accessibility to finance, this book will be particularly valuable to those studying within geography, urban studies and development, as well as to practitioners in NGOs and international institutions.

Kavita Datta and **Gareth A. Jones** are lecturers in Geography at Queen Mary and Westfield College, London and University of Wales, Swansea respectively.

ROUTLEDGE STUDIES IN DEVELOPMENT AND SOCIETY

HOUSING AND FINANCE IN DEVELOPING COUNTRIES

*Edited by Kavita Datta
and Gareth A. Jones*

London and New York

First published 1999
by Routledge
11 New Fetter Lane, London EC4P 4EE

Simultaneously published in the USA and Canada
by Routledge
29 West 35th Street, New York, NY 10001

Typeset in Garamond by M Rules
Printed and bound in Great Britain by Biddles Ltd,
Guildford and King's Lynn

British Library Cataloguing in Publication Data
A catalogue record for this book is available from the British Library

Library of Congress Cataloging in Publication Data
Housing and finance in developing countries / edited by Kavita Datta and
Gareth A. Jones.
(Routledge studies in development and society)
Includes bibliographical references and index.
1. Housing – Developing countries – Finance – Case studies.   2. Housing
policy – Developing countries – Case studies.   I. Datta, Kavita.   II. Jones,
Gareth A.   III. Series.
HD7391.H667   1998
363.5'8'091724 – dc21   98–24528

ISBN 0 415 17242 X

TO PRIYA AND HUW

CONTENTS

CONTENTS

FIGURES

TABLES

CONTRIBUTORS

Kwame Addae-Dapaah is lecturer at the School of Building and Estate management at the National University of Singapore. He obtained his first degree at Kumasi, a Masters at the University of Reading and his doctorate at the University of Strathclyde (UK). He has published widely on housing finance, most recently in *Habitat International*.

Alana Albee is a freelance consultant and has worked on the design, management and evaluation of micro-finance schemes in Sri Lanka, Vietnam, Cambodia, Sudan, Thailand and Bangladesh. She has been a lead consultant for various United Nations agencies to develop methodologies for determining the impact of micro-finance at the household level.

Robert-Jan Baken is a sociologist associated with the Department of Sociology of Development, Vrije Universiteit, Amsterdam, and the Institute for Housing and Urban Development Studies, Rotterdam. His research is in urban land and housing issues, principally in Pakistan and India where he has conducted fieldwork since 1988. He is currently completing his doctoral research into the allocation land across income groups and the role of government in Andhra Pradesh (1971–93). Robert-Jan is co-author (with Jan van der Linden) of *Land Delivery for Low Income Groups in Third World Cities* (Avebury) and has published numerous papers and produced two videos on urban issues.

Joel Bolnick After several years in exile in the USA (1977–85), Joel returned to southern Africa to work on rural development in Zimbabwe and Botswana. He then spent two and a half years as a lecturer in African Politics at the University of Cape Town. Disgruntled with government development programmes and high-brow academics, he started looking for more challenging and meaningful work. A chance contact resulted in an offer to organize the Broedstroom Meeting on Land and Shelter, since when Joel has been one of two directors of the People's Dialogue; the other is Ms Iris Namo from Kathlehong who runs the resource centre in Johannesburg.

Robert M. Buckley is Principal Economist at the World Bank, where he has worked since 1986. He previously served as Chief Economist at the US Department of Housing and Urban Development, as Visiting Economist at the Department of the Environment, UK, and as Professor-researcher at Johns Hopkins University, the Urban Institute, Syracuse University and the University of Pennsylvania. He has worked in over twenty five developing countries and is author of *Housing Finance in Developing Countries* (Macmillan).

Serena Cosgrove has a Masters in Anthropology from Northeastern University, Boston, where she has just completed her doctoral research entitled 'Engendering credit: a comparison of two models of micro-enterprise lending in El Salvador'. She has conducted consultancies for the International Rescue Committee, the Share Foundation and the World Health Organization. She has published in the *Journal of Learning and Advocacy* and *On the Issues: The Progressive Women's Quarterly*. Serena currently holds fellowships from Fulbright and the Interamerican Foundation. She has lived in Central America since 1986.

Kavita Datta is lecturer in the Department of Geography, Queen Mary and Westfield College, London. Her research interests are the gendered nature of urban housing markets; the organization and performance of low-income rental and sharing markets, and gender-specific migration patterns. She has had articles published in *Habitat International*, *Cities* and the *International Journal of Population Studies*.

Beverley L. Drought is a doctoral candidate in the Department of Geography at the University of Wales, Swansea, where she is writing her dissertation on 'Resettlement in the Russian Federation: a case study of refugees and forced migrants from Tadzhikistan'.

Nandasiri Gamage is a community activist from a low-income community in Colombo, Sri Lanka. He is a founding member of the Community Agent Service (Praja Sahayaka Sewaya). A gifted story-teller with a strong commitment to reflecting the daily life and concerns of people in low-income areas, he has travelled to India, Thailand, Vietnam, Denmark and the UK to share his experiences.

Ann Gordon is an agricultural economist with the Natural Resources Institute (UK), currently seconded to the International Service for National Agricultural Research (Consultative Group for International Agricultural Research) in The Hague. Most of her work is concerned with development and change in agricultural markets in Sub-Saharan Africa, Asia and Latin America. The role of food production, processing and trade in women's livelihoods is an important dimension of this work.

Katherine V. Gough is a British geographer and is currently employed as assistant professor in the Department of Geography, University of Copenhagen. Her main research interests are in sustainable urbanization in the Third World focusing in particular on low-income housing in Latin America, particularly Colombia, and West Africa. She is currently engaged in a research project on the socio-economic and environmental consequences of urban growth in peri-urban Accra, Ghana. She has published articles in a range of international journals.

Gareth A. Jones is lecturer in the Department of Geography, University of Wales, Swansea. He obtained his doctorate from the University of Cambridge and held a CONACYT scholarship at the Universidad Autónoma Metropolitana-Azcapotzalco, Mexico City. He is co-editor of *Methodology for Land and Housing Market Analysis* (UCL Press) and of *Dismantling the Mexican State?* (Macmillan) and has worked extensively on urban issues in Mexico, Colombia, Ecuador and South Africa. In 1998 he was Researcher-in-Residence at the Center for US–Mexican Studies at the University of California-San Diego.

Thomas Klak is Associate Professor of Geography and Director of the Latin American Studies Program at Miami University in Oxford, Ohio, and Adjunct Associate Professor of Geography at Ohio State University in Columbus, Ohio. Over the last seven years he has published more than twenty articles and book chapters concerning development issues and state policies in Latin America and the Caribbean. He is currently editing a book entitled *Globalization and Neoliberalism: The Caribbean Context* (Rowman & Littlefield).

Diana Mitlin is an economist working in the Human Settlements Programme at the International Institute for Environment and Development (IIED), London, where her interests include NGOs, urban community development, sustainable development, housing finance and the use of participatory methodologies in urban areas. She is a contributor to the UNCHS's *An Urbanizing World: The Global Report on Human Settlements*, the project leader of the ODA consultancy 'Local Initiatives Facility for the Urban Environment' and co-ordinator of an IIED research project on community environmental initiatives in Karachi and Manila. Her publications include co-authorship of *Funding Community Initiatives* and *Environmental Problems in Third World Cities* (both Earthscan). Diana is managing editor of *Environment and Urbanization*, chair of the UK charity Homeless International and a trustee of Intermediate Technology.

Eduardo López Moreno is Co-ordinator of the Centro de Estudios Metropolitanos at the Universidad de Guadalajara, Mexico. He holds a doctorate in urban sociology from the University of Paris IV and is author of two important books on social housing and colonial urban design. He has held consultancies on project evaluation and housing policy in Haiti, Angola, Vietnam and Brazil. His research interests include social housing and the integration of the popular sector into productive modes of self-production and control of local communities.

Sheela Patel is a co-founder and director of the Society for the Promotion of Area Resource Centres (SPARC) which was set up in 1984. Sheela has worked on issues of women and urban poverty since 1974 when she graduated from the Tata Institute of Social Sciences in Mumbai. She has a particular interest in the study of those institutional arrangements which are owned by the poor and which allow them to gain access to social and economic environments which are otherwise denied them.

William J. Siembieda is Professor and Director of the Department of City and Regional Planning, College of Architecture and Environmental Design, California Polytechnic State University, San Luis Obispo. He has been visiting professor at universities in Brazil and Mexico and held consultancies for the governments of Chile, Colombia, Mexico and Cuba. His research interests include planning for large-scale projects, and urban land economics, land pricing, and the application of Geographical Information Systems to town planning and land supply policy.

Peer Smets is a Research Associate at the Department of Cultural Anthropology and Development Sociology of the Vrije Universiteit, Amsterdam, where he is currently completing his doctorate on housing finance for low-income households in Hyderabad, India. He has published numerous articles on government bureaucracy, financial self-help organizations, housing finance and slum development in India and South Africa.

Marlene Smith holds undergraduate degrees from the College of Arts, Science and Technology in Kingston, Jamaica, and from Rutgers University in New Jersey, as well a Master's degree in Urban Studies and Planning from the Massachusetts Institute of Technology. She has worked as a planner for the Petroleum Corporation of Jamaica and a research officer for Jamaica's Urban Development Corporation. Since 1991, she has worked as a Senior Research Officer at Jamaica's National Housing Trust.

PREFACE

OVERVIEW OF GENERAL THEMES

About half of the world's urban population lives in poverty and about 600–800 million people occupy substandard housing. Even so, housing investment represents 3–10 per cent of GDP in most developing countries; it is the largest item of non-food household expenditure and the most valuable asset possessed by most low-income households. In such circumstances it is almost a truism that one solution to poor housing conditions is to raise the amount of finance available to low-income households so as to improve the means by which they are able to build more and better quality housing. That said, there is remarkably little comparable information on housing markets and housing finance. As a United Nations Centre for Human Settlements report points out,

> That international agencies have as yet been unable to mobilize the funds and elicit the cooperation of member countries in collecting and supplying information is . . . indicative of the lack of organization in the sector and of the priority attached to housing by policy-makers.
>
> (UNCHS 1991b: 3)

Even events such as the debt crisis and the demise of public housing finance programmes have prompted a slow response by many governments and multilateral institutions. Yet the argument that correctly structured finance systems and governments which adopt 'enabling' strategies, including deregulation of the financial sector, can deliver improved housing is beginning to gain purchase (Malpezzi 1990, van Huyck 1987, World Bank 1993b). Indeed, one observer has argued that housing finance has gone 'from being something of a policy carbuncle, a mere conduit to carry out government policies, into an issue of major concern' (Buckley 1996: x).

Understanding the operation of, and potential for, housing finance is important since in many developing countries 'housing' policy is about establishing new and more innovative finance policies. Governments have had to scale down or abandon programmes of mass house construction and sites-and-service schemes are rarely talked about in terms other than as an interesting experiment. Moreover, while the early attempts to establish state-run banks and mortgage programmes did raise the amount of capital devoted to the housing sector, most schemes operated with substantial subsidies, targeted already 'privileged' groups and failed to reach the poorest households, ignored non-conventional household structures and had a strong bias against women. One response has been to deregulate the financial sector in order to increase the participation of the private sector and the amount of capital investment in housing. Nevertheless it is apparent that most financial institutions have been reluctant to become involved in housing or the low-income housing market more specifically.

Compared to the poor performance of the public sector, informal methods of housing finance appear to be making a substantial contribution toward meeting low-income needs. Moreover, informal finance seems to possess many of the qualities which advocates of self-help recommended for housing strategies. Conversely, when programmes have not been based on the flexibility that low-income households require to finance their housing, or have been too rigid in attempting to capture the principal features of self-help in official programmes, they have often produced additional costs for the poor. In short (to corrupt the title of one of the most influential housing texts), along with 'freedom to build', low-income households need the 'money to build'.[1]

Increasingly, therefore, the rhetoric if not always the reality is to strengthen informal finance systems, usually with the assistance of the NGO sector. Informed by the perceived success of innovative schemes such as those of the Grameen Bank and micro-finance to support the establishment of small enterprises, many analysts have begun to consider whether similar models might be appropriate for housing.[2] Yet, there remain significant conceptual and logistical problems with translating this potential into practice. Indeed, the *Declaration and Plan of Action* agreed at the Microcredit Summit held in Washington in 1997, which set out to highlight the achievements and transferability of micro-finance principles, afforded almost no attention to housing, or to the provision of urban services or land.[3] There is an urgent need to examine the potential for, and constraints on, micro-finance programmes and informal schemes delivering finance for housing purposes.

One approach to is to harness the estimated 10,000–20,000 NGOs working in developing countries, even though very few have direct experience of housing or housing finance.[4] Nevertheless, one finds that NGOs are involved in many of the most innovative finance schemes, through funding community initiatives, as direct participants or as the providers of expertise. The perceived success of many of these NGOs has caught the imagination of national governments and international agencies, notably at the Istanbul Habitat II conference in 1996, and many schemes are under pressure to scale up their activities. Perhaps the most strongly advocated route is for NGOs to adopt finance-only or minimalist strategies, by dropping their welfare or social development focus, and to concentrate on delivering finance without recourse to recurrent subsidies.

One contentious area of debate between the 'evangelists' advocating a minimalist approach and NGOs struggling to retain their social development remits is the importance of gender as a critical component of finance programmes.[5] The participation of women as savers, borrowers and managers in micro-finance programmes, whether by accident or design, appears to be fundamental to repayment rates and is well documented.[6] This has led NGOs and networks such as Women's World Banking and the International Coalition on Women and Credit to advocate the inclusion of women and (other) disempowered groups into micro-finance programmes. Moreover, a growing number of institutions have sought to use micro-finance as a means of 'reaching poor women', who make up 60–70 per cent of the world's poor, in 'gender-sensitive' ways and to 'empower' them in the process.

Yet, there is debate as to whether the empowerment of women can be achieved through their participation in finance programmes, especially without the addition of training and advice. While informal and NGO finance programmes provide support for women in ways not directly related to finance itself, for example in establishing women-only networks, studies are beginning to reveal that a woman's income may disappear into the 'household' budget and that micro-enterprise schemes tend to confirm rather than challenge gender stereotypes. Indeed, the perception of women as 'good managers' which is inherent in many micro-finance programmes may serve to further disempower women by adding extra burdens to their load, serving as a conduit for partners or other household members to increase their access to resources, and make them vulnerable to social pressure from lending agents or community leaders. The key, therefore, is the form which finance delivery takes and the nature of the organisational support.

Nowhere, perhaps, are these issues better illustrated than in the gendered provision and design of housing in developing countries, which has been traditionally gender-blind in both the conception and the execution of programmes.[7] Thus, men and women continue to experience differential and unequal access to housing, productive and reproductive work within the house, and control over the housing environment and its resources. There is also evidence that possession of housing is an important element in survival strategies for low-income households, particularly those headed by women. In a sense, then, housing is a vital form of saving which, with the commodification of urban housing markets, is likely to increase in value over time.

Finally, establishing micro- or housing finance schemes in developing countries does not take place in a political vacuum. Although most studies take the political context for granted, this is a mistake. Governments are increasingly drawing links between their macro-economic policies and housing finance, as part of structural adjustment programmes or broader reform packages that require conducive political environments. In some instances, many of the agents of micro- or housing finance are themselves directly involved in a 'political' process, either by working with government directly, as in the case of many NGOs, or by advocating changes to policy. In addition, political events have been the incentive for the initiation of finance programmes, as in El Salvador and Russia, or have dramatically changed the form which they have taken, as in South Africa.

EDITORIAL MOTIVATIONS

There are two principal personal motivations for compiling this volume. The first stems from the observation that there is no single source of papers on housing finance available to a wide and inter-disciplinary readership, and few volumes have captured the diversity of the experiences with housing finance. Indeed, the publications which are available restrict their optic to exclusive aspects of housing finance such as economic theory or the role of NGOs, and rarely consider a fuller range of approaches or issues. Reflecting this narrow scope, perhaps, much of the available information is limited to workshop and internal reports, or official publications. Indeed, despite much talk of mainstreaming of best practice,

> there has been very little exchange of experience and learning about housing finance for low-income groups initiated within the NGO sector. . . . Well-documented case studies are very scarce, and those that do exist rarely focus on the details of financial management and community organisation.
>
> (Mitlin 1997: 30)

We were keen, therefore, to put together a volume which covered the experiences of academics and NGOs working on themes of housing, finance and gender, and we are particularly pleased that the invited contributions include those of representatives of NGOs and low-income communities in developing countries. We were also aware that many texts have tended to concentrate upon either formal or informal finance, despite evidence showing the linkages between the two, and to concentrate upon mechanisms for delivery rather than the effects on housing, household structure and gender relations. Again, we have attempted to stretch the canvas and include as many angles on housing finance as is possible in an edited collection.[8]

The second motivation was provided by a visit to South Africa to look at housing finance. Our research there was intended to appraise the rapid political and economic reforms with specific reference to the role of NGOs and to their impacts at the household level. During our visit, however, we were struck by the inconsistent ways in which different agents were using finance as a means to build or improve housing. Among a host of very positive memories from that trip, two meetings illustrate this point.

Late one afternoon, we were invited to the weekly meeting of a credit rotation scheme in a large squatter settlement. The scheme was managed by a locally based NGO, had been operating for a number of years and had upwards of fifty women members. At the meeting we observed the pride with which the women talked about the running of the scheme and how it had taught them to manage not only finance, but people. The women talked passionately and convincingly about the scheme, their hopes for the settlement and their families. They were keen to show us how the group kept detailed records, that none of those present had ever fallen behind in making their weekly contribution to the general fund and had never defaulted on the small loans made from time to time to cover emergencies. One could only feel impressed by the feeling of solidarity in the room and, to put the event into academic terms, a palpable sense of empowerment. Our only misgiving was when, in answer to a question about how many houses had been built by the scheme, the reply was five. As the scheme was making long-term loans with the number dependent upon the level of group savings and a government subsidy, a mental calculation suggested that the last member would have to wait fifteen years to complete a house.

At the other extreme, a few days later we had a discussion with a mortgage finance company which, so their publicity claimed, specialized in finance for low-income households. The discussion was technical and frank, the respondents well aware of the need to charge market rates, the problems of valuing squatter housing as collateral, phasing repayment schedules and establishing insurance contracts to cover insecure incomes. Clinging to the memory of undergraduate economics courses, we just about kept up with the jargon while wondering how low-income households were going to understand these concepts: housing finance had become a science. In contrast to the members of the credit rotation scheme who talked about people, here the conversation was quantified in terms of internal rates of return, gearing and portfolio performance indicators. The respondents admitted that their definition of low-income did not reach down to the very poor, but when asked about how many houses they expected to fund in the next few years, the reply was in the thousands.

These two meetings captured some of the difficulties in our current understanding of and approaches to housing finance. They illustrate the wide variety of experiences of housing finance, which range from the technical, finance-based approaches, to community banking and informal schemes. They also illustrate the different perspectives of those who regard terms such as 'sustainability' from a purely economic point of view and who share a concern with the replicability of the finance scheme, and others who use a broader definition to include community solidarity and the need to ensure the sustainability of the households' resources.

DEFINITION OF TERMS

Given the multi-disciplinary nature of this volume and the diverse backgrounds of the contributors, it has not always been possible to impose a strict rigour of terminology. As a general rule, we use the term micro-finance to denote small-scale savings, and the giving of loans, grants and subsidies, in preference to the more popular 'micro-credit' which implies just one side of the financial transaction (making loans) even though it is rarely used in this limited format.

We are also aware of the difficulties with using terms such as 'formal' and 'informal', particularly where these describe a form of financial organization (for example, an informal savings scheme) which is likely to deposit funds in a formal institution. Equally, many indigenous NGOs are informal in the sense that they are not registered with the government, or at least not for the purpose of conducting financial transactions, but are likely to be regulated by internal codes, to be subject to fiscal controls (if only by why of exemption) and to not undertake illegal activities. As with informal/formal distinctions identified in other areas of development studies, the creation of such categories creates problems for cross-national comparison. Chit funds, for example, are unregulated in some countries and might be classified as informal but are regulated in some others. Given that the definition of 'informal' and 'formal' is locally determined we have not sought to impose a rigid editorial template over authors' use of such terminology.

One of the problems of the existing literature is the wide variation of meanings attached to 'housing'. This has important consequences for how we interpret all studies as well as those presented in this volume. The Grameen Bank's housing loans, for example, are sufficient to buy four concrete pillars, 21 square metres of roof sheeting and a low-cost sanitary toilet. In some societies, the notion of housing as a set of materials would be unacceptable, while in others this combination would fall short of a reasonable attempt at improvement. Similarly, lending for housing 'improvement' would not constitute housing finance for some agencies, as it does not provide housing *per se* (that is, new construction with full amenities).

Partly related to these problems of definition associated with notions of housing is the conflation of a range of financial mechanisms under the rubric of 'housing finance'. The range of mechanisms is truly impressive but, broadly, involves the provision of loans, subsidies, the organized mobilization of savings, or any combination of these. While the diversity of programmes is a healthy indication of innovation by households and NGOs in difficult circumstances, it makes comparison impossible. Indeed, many studies prefer to use rather nebulous measurements such as 'interventions' or 'loans' as a measurement of impact and are rather quiet about how many houses have actually been built or improved. Reading off an assumed level of housing provision or improvement from the number of loans distributed is hazardous and would require more universal terms of reference than we could impose in an edited volume. Again, therefore, we have allowed authors considerable latitude in defining what constitutes 'housing finance' according to their own experience.

We hope that this volume will provide a useful reference point for academics, policy-makers, NGOs and finance organizations, in both the formal and the informal sectors, such that each can learn from the experience of the other. Improving our knowledge, however, must serve to assist those low-income households who have demonstrated their ability to construct housing under the most difficult of financial circumstances.

Kavita Datta and Gareth A. Jones
Swansea, UK
January 1998

NOTES

1 Turner and Fichter, *Freedom to Build: Dweller Control of the Housing Process* (1972), New York: Macmillan.

2 For excellent discussions of micro-finance see Adams and Fitchett (1992), Bouman and Hospes (1994), Hulme and Mosley (1996), Johnson and Rogaly (1997) and vol. 8, no. 2 of *Journal of International Development* (1996).

3 See the *Declaration and Plan of Action* and *The Microcredit Summit Report* (The Microcredit Summit 1997a, 1997b). These and the other documents adopt an implicit anti-urban and housing stance, referring to beneficiaries as living in 'villages' and promoting finance for enterprises rather than construction or service improvement. Most of the NGOs and programmes presented at the Summit either had a specifically rural focus or concentrated on enterprise or broad welfare-related issues. The target set at the Summit for the multilateral development community to provide US$ 21.6 billion to assist 100 million borrowers by the year 2005 (an average loan of about US$ 150) would be insufficient to finance any substantial improvement in housing conditions.

4 We are certain that the figure quoted here would be disputed by many and with good reason. Many NGOs operate without official recognition or with inconsistent definition (not-for-profit, non-profit making, etc.), and through umbrella organisations, which makes assessment of their number difficult. We recommend Bratton (1989, 1990), Carroll (1992) and Edwards and Hulme (1992a) for general discussions of NGOs in development.

5 The term 'evangelist' is Rogaly's (1996).

6 For discussion of the relevance of micro-finance to women's income generation see Ackerly (1995), Ardener and Burman (1995), Berger (1989), Goetz and Gupta (1996) and Smillie (1995).

7 For a review of the importance of gender to housing and household see Chant (1997), Dandekar (1992), Moser (1992b), Moser and Peake (1987) and Varley (1994, 1996).

8 We are, of course, aware that the contributions to this volume leave a number of important themes untouched: not least, the importance of finance to the rental sector, the relationship between housing and municipal finance, and cultural constraints upon finance in, for example, Islamic societies.

ACKNOWLEDGEMENTS

The editors owe a special debt to a number of people and institutions. At Routledge, Casey Mein and Sarah Carty are owed particular thanks for their patience as firm deadlines became moving targets. The ODA (now DfID) provided a grant to conduct research on 'The Impact of Finance Reform on Housing Markets in Botswana and South Africa' (ESCOR R6376).

An earlier version of Robert M. Buckley's chapter appeared as chapters 11 and 12 in *Housing Finance in Developing Countries* (1996) published by Macmillan. Some of the material presented in the chapter by Katherine V. Gough is reprinted from *Habitat International*, vol. 20, Katherine V. Gough, 'Self-help housing in urban Colombia: alternatives for the production and distribution of building materials', 635–51 (1996) with kind permission from Elsevier Science Ltd, The Boulevard, Langford Lane, Kidlington OX5 1GB. Material presented in the chapter by Kwame Addae-Dapaah is reprinted from Habitat International, Vol.20. K. Addae-Dapaah and K.M. Leong 'Housing finance for the ageing Singapore population: the potential of the home equity conversion scheme', 109–20 (1996) and vol. 20. K. Addae-Dapaah and K.M. Leong 'Housing finance for the ageing Singapore population: the potential of creative housing finance schemes', 625–34 (1996) also with kind permission of Elsevier Science Ltd.

Part I

POSITION PAPERS

1

FROM SELF-HELP TO
SELF-FINANCE

The changing focus of urban research and policy

Gareth A. Jones and Kavita Datta
University of Wales, Swansea

INTRODUCTION

A strong relationship between levels of urbanization and wealth has been demonstrated both theoretically and empirically in numerous studies (Malpezzi 1990, World Bank 1993b). Yet, in practice, it has proven to be very difficult to harness the enormous wealth generated through urbanization for the benefit of low-income households. Traditionally, faced with other development priorities, governments and international agencies have been reluctant to encourage investment in housing, which has often been seen as an item of consumption (UNCHS 1991b, 1996). Moreover, many of the first wave of housing finance institutions were poorly managed and contributed to macro-economic disruption (Buckley 1996, Buckley and Mayo 1989). Even by the late 1980s, Renaud was able to observe that 'few aspects of economic development remain as unexplored and poorly analysed as the potential to induce financial development and ways to improve the financing of housing' (1987b: 30).

These practical and conceptual difficulties notwithstanding, during the 1990s housing finance moved to the top of the urban agenda (World Bank 1993a). Under pressure to reform urban management, governments have made important legislative and institutional reforms to enable private institutions and non-governmental organizations (NGOs) to have a greater role in the provision of housing finance.[1] The lead of the World Bank has been especially important in making the shift from housing projects towards the delivery of housing finance (World Bank 1993a). From 1983 to 1988, Bank lending for housing finance exceeded the total for sites-and-services from 1972 to 1988 and by 1989 almost one-half of all Bank urban lending was for housing finance programmes (Buckley this volume). This reorientation went beyond the need to deliver more and better housing, to make urban policy compatible with macro-economic management, particularly in the context of structural adjustment programmes in which control of foreign exchange risks and fiscal probity have been paramount (Buckley 1996, Jones and Ward 1994, Malpezzi 1994).

While this policy shift has generated a significant literature on the supply-side of housing finance, our understanding of the uses to which households have put finance and the implications for house construction is less advanced and requires attention to a more disparate literature. The aim of this chapter is to draw together the emergence of leading-edge themes of housing finance in developing countries and to place these within the broad context of the housing problem and the evolution of housing finance policy and research. To some extent, our aim is also to play devil's advocate to some of our own contributors and the literature in general. Following the policy debates on housing finance prior to Habitat II in Istanbul (1996) and on micro-finance in the lead up to the Microcredit Summit in Washington (1997) we noted the high degree of optimism and the tendency to cite 'positive' experiences. General claims have been made for the ability of private-sector institutions to go down-market and lend to lower middle-income and, with less regularity, low-income households. Partly building upon the experience of programmes for micro-enterprise and consumption finance, arguments have been made for the ability of NGOs to deliver housing finance. The extension of micro- and housing finance was also being linked to the empowerment of low-income households and communities, and especially of women. While not wishing to dispute the validity of specific cases, our argument is for caution and to concur with the stoic advice of Adams and von Pischke that 'debt is not an effective tool for helping most poor people enhance their economic condition' (1992: 1468). We believe that the potential of housing finance as a means to improve living conditions can be better understood if governments, NGOs and international agencies analyse the weaknesses as well as the successes of recent experience.

HOUSING AND FINANCE

The scale of the problem

In developed countries, housing conventionally costs three to four times the combined annual incomes of the owners so that virtually all housing is bought or built with finance. With a typical deposit of between 5 and 20 per cent of the total price, finance is needed to cover about 80 per cent of costs and is usually repayable over twenty years. For most borrowers this debt burden is manageable because, although the finance is repaid at positive interest rates, income growth is likely to be relatively constant and predictable, and the family cycle dictates that most households will have reduced outgoings over time. Although not affecting all households equally, positive interest rates in the short term are partly offset by returns on savings and, in the event that incomes do not increase or that savings are unable to smooth the shortfall, data show that house values tend to outperform inflation. Should normal financing arrangements be insufficient to cover remaining balances of principal, households can realize insurance policies (which can also cover loss of earnings or excessive interest rate increases) or expect an equity windfall from inheritance. In a worse case scenario, lenders are accustomed to offering a range of refinancing packages such as lengthening the loan period or allowing interest rate holidays as a matter of course and often at little additional cost (Diamond and Lea 1992, Hamnett 1994).

By contrast, households in developing countries face a series of problems in attempting to access finance with which to resolve their housing needs Despite enormous absolute housing deficits and the need to improve the existing stock, housing finance often represents less than 10 per cent of all financial transactions. To make matters worse, many housing finance institutions possess losses amounting to many times the value of their capital reserves, have been prone to invest in highly speculative ventures with consequent boom-bust swings in their portfolio and have a track record of delivering funds only when government subsidies are available and only then to the better-off 10–20 per cent of households (Boleat 1987, UNCHS 1991b). Not surprisingly, most estimates of housing investment as a proportion of GDP in developing countries provide figures substantially below those of developed countries (Buckley 1996, Malpezzi 1990, Renaud 1987b, World Bank 1993b).

This situation means that most households have to 'chase loans' with upwards of 80 per cent of housing finance transactions taking place in the informal economy (Renaud, 1987b: 184, Okpala 1994). The size and the nature of the gap between supply and demand is indicated by Merrett and Russell

> The UNDP suggests that in the world of conventional finance, the minimum loan size may be large, whilst the poor require small loans. The formal loan period may be for many years, whilst the low-income household seeks a short maturity, and the availability of follow-on funds. Banks require regular repayments of principal, but workers with insecure earnings need flexible loan schedules. The formal system may offer money only for completed dwellings, when what is needed is money for the initial stages of self-help or improvement of the self-help house once it is constructed. Finance may be offered with conditions written in a language of considerable sophistication, when those in housing poverty require a legal agreement which is

easily understood, and under which payments are possible in local premises open throughout the working day for cash payments.

(1994: 60)

To this salutary list one might add that formal finance institutions are rarely willing to assist with the purchase of land, especially where the tenure is insecure, to provide assistance with improvements to the rental housing stock or to support non-conventional household arrangements such as sharing or multiple-family compounds. These limitations have implicit gendered conse-quences as rental and shared housing are of particular importance to low-income women who often lack the means to become home-owners (Chant 1997, Datta 1995, Miraftab 1994). Of course, some of these problems are derived from broader inequalities and do not derive directly from the short-comings of finance institutions. As a number of authors point out, weaknesses in the delivery of housing, the lack of affordable land and an inefficient urban administration are as much causes of the poor financial condition of the urban sector (Rakodi 1995, Renaud 1987b, Struyk and Turner 1986).

The scale and direction of some of these problems have been captured by the Housing Indicators Program (World Bank 1993b).[2] Table 1.1 provides details of the basic housing conditions for countries featured in this volume and for some of those most frequently mentioned in the literature. The details show high levels of urbanization for countries in Latin America and the Caribbean as well as significant levels in South Africa and Malaysia. Squatter housing (illegal land occupation) is clearly a major problem in

Table 1.1 Housing conditions in developing countries

	GNP per capita (US$)	Level of urbanization (%)	Squatter housing (%)	Unauthorized housing (%)
Bangladesh	210	16	10	78
India	350	27	17	48
Ghana	390	33	0	40
Indonesia	570	31	3	70
Colombia	1,260	70	8	8
Thailand	1,420	23	3	17
Jamaica	1,500	52	33	50
Chile	1,940	86	0	20
Malaysia	2,320	43	12	12
Mexico	2,490	73	4	16
South Africa	2,530	60	22	34
Singapore	11,160	100	1	1
Developed countries	14,000-plus	—	—	—

Note: The Indicators Program uses key cities (usually capitals) to represent 'countries'. The cities are Dhaka, New Delhi, Accra, Jakarta, Bogotá, Bangkok, Kingston, Santiago, Kuala Lumpur, Monterrey, Johannesburg and Singapore. Income is converted to Purchasing Power Parity.

Jamaica and South Africa, although recent figures for both Colombia (Bogotá) and major Mexican cities provide figures many times higher than those of Table 1.1. Unauthorized housing is a universal problem, accounting for about 30 to 60 per cent of housing. Although differences in the ability of low-income households to acquire finance alone are unlikely to explain cross-country differences in housing conditions, the data indicate that while average income in some countries might be greater the availability and type of finance are of critical importance. For example, in India where few low-income households have access to formal finance the proportion of households living in permanent structures is significantly less than Mexico which devotes more resources to housing and has a range of delivery institutions (for a comparison of Korea and the Philippines see Struyk and Turner 1986).

A more detailed look at housing finance conditions is provided by Table 1.2. The house price/income ratio is credited as a proxy for the level of dysfunction in the housing market: a high ratio indicates restrictions to supply and a low ratio indicates insecurity of tenure or poor-quality accommodation.[3] The data reveal lower ratios for Africa and Latin America (but not Colombia) than for South Asia, the Middle East and North Africa (World Bank 1993b: 11–12). The ratio also indicates whether better housing finance delivery will improve housing conditions: if the relative cost of housing to income is high, more resources will make little impact upon housing conditions unless institutions are willing to lend many times over conventional income-value ratios. As the column for credit/value ratio indicates, for most countries this is unlikely to be the case.

Table 1.2 also shows that financial institutions seem unable to supply adequate resources for housing purposes. The housing credit portfolio is the ratio

Table 1.2 Housing finance in developing countries

	House price/ income ratio	Credit/value ratio	Housing credit portfolio
Bangladesh	6.3	1.8	3.0
India	7.7	13.8	2.4
Ghana	2.5	—	2.9
Indonesia	3.5	35.1	3.3
Colombia	6.5	60.0	11.0
Thailand	4.1	66.3	7.1
Jamaica	4.9	28.0	22.0
Chile	2.1	44.4	20.4
Malaysia	5.0	73.4	22.0
Mexico	3.7	77.0	18.0
South Africa	1.7	75.0	39.0
Singapore	2.8	41.0	15.0
Developed countries	4.5	90.0	25.0

of total mortgage loans to all outstanding loans in the commercial and government sectors. It is thus a measure of the relative size of the housing finance sector. A low figure indicates either an underdeveloped sector (perhaps because of opportunities elsewhere) or constraints imposed on the sector by government.[4] Although this indicator only includes finance in the formal sector, it demonstrates that housing finance represents only a low proportion of financial transactions in nearly all developing countries and especially those with the lowest levels of GNP.[5]

Of course, it is extremely difficult to establish housing finance institutions under any circumstances, but in developing countries the problems can be particularly acute. First, the very nature of housing finance means that institutions are required to borrow short and lend long, thereby extending the conventional period of time before sustainable operations can be reached. Second, in economies with a shortage of capital the competition for funds is high and institutions may have to pay premiums in order to acquire capital, putting further pressure on their portfolio. Until the 1980s, most governments attempted to resolve these problems through the creation of designated public housing and finance institutions.

The rise and fall of public-sector housing finance institutions

While many of these institutions assisted in the construction of large numbers of housing units, almost all have encountered severe financial problems and many have been wound up, privatized or obliged to take on private partners (Munjee 1994, Pugh 1994). One reason for this poor performance relates to the methods used to fund these institutions. Their reliance on the seemingly endless supply of funds from (sometimes nonredeemable) payroll levies, central government transfers or the enforced sale of government-backed stock to other public departments removed the pressure to be efficient (Boleat 1987, Buckley 1996, Rahman 1994, USAID 1997).[6] In a number of countries, funds were the result of foreign loans which were to be repaid out of general government expenditure without the housing finance institution being accountable to the donor for their use or providing adequate details of repayment and subsidy rates (Rahman 1994, Valença 1992). In general, interest rates were often adjusted only periodically and were not subject to central bank control, monetary correction or the indexation of balances, aiding the steady path toward bankruptcy (Munjee 1994, UNCHS 1991b). The constant resource top-ups also removed any desire to innovate by developing new financial instruments, look for new client groups and offer bespoke finance packages. Nevertheless, some institutions expanded their remit to become housing and land developers without providing much evidence for having performed particularly well at their principal role.

8

A second characteristic of the public finance institutions has been a poor organizational record. Valença (1992: 39) summarizes the conditions of Brazil's housing finance system by the 1980s as one of 'crisis, chaos and apathy'. Notoriously inadequate fund collection and loan enforcement rates exemplified these conditions. In Chile 60 per cent of loan recipients through the Basic Housing Programme were in arrears despite discounted interest rates, and about 99 per cent of loans given by the House Building Finance Corporation in Pakistan were in default although record-keeping was so poor that the precise figure is not known (Buckley and Mayo 1989, Malik 1994, Rojas and Greene 1995). Alarmingly, default rates were high even from households which had received large (if rarely transparent) levels of subsidy and despite some attempts to index repayments to wages (which were kept artificially low) rather than consumer prices, which were increasing at a faster rate (Buckley 1996).

These conditions were accentuated by political manipulation that passed institutions from one ministry to another at short intervals (Okpala 1994: 1577–8, Valença 1992). Institutions also became a part of the 'patronage bureaucracy' serving sectoral interests to give favourable treatment to labour groups or to co-opt opponents (Bond 1990, Klak and Smith, Siembieda and López this volume). In Ecuador, for example, employees of the Housing Bank were entitled to an annual mortgage subsidy three times the level of GDP per capita in addition to other generous benefits (Klak 1993: 667). Having brought some of these groups into housing finance more as 'vote banks' than as clients there was a natural tendency to waive loan repayments and provide unconditional amnesties to defaulters at election time. In Sri Lanka, the Million Houses Programme was hijacked by political interests and eventually operated with only one-half of loan recipients making repayments on time or of the correct amount (Sirivardana 1986).

In the short term, this situation appeared sustainable so long as economic growth, low inflation, and stable exchange rates and commodity prices supported the large capital injections to cover subsidies and administrative inefficiency.[7] However, as the economic crisis of the 1980s deepened, the fall in the real value of payroll deductions with rising unemployment, the diversion of revenue sources to fund higher priority areas of the government budget and the withdrawal of savings from negative interest rate bearing accounts left many institutions short of capital (Klak 1993, Rakodi 1995, Valença 1992). Between 1979 and 1989, assets per capita in financial institutions were held at very low levels and even fell, especially from 1985 to 1989 (UNCHS 1991b:12–14).[8] The response was to ration lending and to depart from any pretence at social progression, by going up-market in the hope of finding borrowers with sufficient income to maintain payments (Buckley and Mayo 1989, Okpala 1994: 1581, Valença 1992).

The declining effectiveness of housing finance institutions, coupled with economic and fiscal crises, have made governments more aware of the need to promote savings, reduce subsidies and mobilize domestic resources, and to motivate the involvement of private financial institutions (Buckley 1996, Kim 1997). Many of the most restrictive practices operating in housing finance markets, such as institutional entry requirements and liquidity limits, have been lowered, loan/value ratios made more flexible and a wider definition given to the terms of collateral.[9] The optimistic view was that private institutions would be able to deliver larger quantities of finance more efficiently and with a greater chance for sustainability (Munjee 1994).

Certainly, there is evidence that deregulation and other reforms have promoted greater innovation, notably the development of dual-interest mortgages (DIMs) which allow payments to be indexed so that lenders are not 'taxed' through inflation, secondary mortgage markets, pension-linked housing finance and insurance-backed schemes (Buckley 1996, Kim 1997, Rakodi 1995). One of the best examples is the development in Chile of certificates which are issued with a mortgage, pay interest to the bearer and are traded on the stock exchange. The effect has been to reduce the first-level risk of lenders and capitalize the housing finance market (Buckley 1996). In Thailand, the Government Housing Bank now competes with private-sector institutions for savings and loans by charging commercial rates, provides a range of mortgage instruments and operates with low service charges (Kim 1997: 1601). In Ghana, the government has formed a joint venture with insurance companies and a merchant bank to supply housing finance to middle-income households (Derkyi 1994).

Despite these reforms, there is only tentative evidence of public and private finance institutions moving down-market (Rakodi 1995, Smets 1997, UNCHS 1991b). Banks in Chile have set up micro-finance programmes to reach lower-income households; in the Philippines public housing institutions have provided 'community mortgages' to households employed in the informal sector; and 'apex' institutions in which low-income communities, NGOs and finance institutions are represented have been adopted in Namibia and advocated for Bangladesh (Alonzo 1994, Igel and Srinivas 1996, Lee 1995, Mitlin 1996). But in Pakistan, where new private finance companies have been set up, few lower-income households qualify for loans because the eligibility criteria require proof of five years' full employment, impose a start-up fee equivalent to three months' salary and taxes to approximately 25 per cent of the loan value (Malik 1994). Similarly, in eastern Europe, despite the establishment of DIMs in Poland, building certificates in Russia and indexed credit systems in Bulgaria, the bottom 80 per cent of the income profile has not been reached (USAID 1997). The general picture remains of a highly bureaucratic finance sector which is wedded to the need to establish credible finance arrangements and which places a heavy premium on provision to low-income households due to the perceived level of risk, poor enforcement record and high administration/value ratios.[10]

One move often made by financial institutions to ensure some equity in finance lending has been to establish programmes to channel subsidies to borrowers. Whereas, in the past, subsidies were distributed fairly indiscriminately as part of general programmes, it is increasingly argued that subsidies should be targeted to prevent them being 'captured' by better-off households (Diamond and Lea 1995, UNCHS 1991b). It is argued that the most effective means, both administratively and financially, to target subsidies is through one-off grants rather than recurrent dispersal. Research has also demonstrated that when the source of the subsidy is borrowers (through higher interest rates) or savers (negative interest rates) the effects are particularly harmful to the sustainability of financial institutions (Buckley 1996).[11] In practice, it has proven more difficult to translate these insights into effective programmes. Despite the technical advantages of one-off grants over recurrent subsidies, governments have found it difficult to extricate themselves from the political compromises which led them to favour general subsidies in the first instance (Kim 1997). Moreover, while the use of grants has been adopted in Chile, Colombia and South Africa, bureaucratic logjams and problems establishing tenure security have caused a slow disbursement of subsidies, which have been taken up by better-off low-income households (Gilbert 1997, Rojas and Greene 1995, Tomlinson 1995).

For the most part, research on public and private housing finance institutions has concentrated upon the supply side and not looked in nearly as much detail at how households have used this funding when it has been made available. Data show institutions favouring households on above-average incomes, employed in the formal sector and where the household 'head' is male (Klak and Smith this volume, Struyk and Turner 1986). There is also some indication that households have used formal finance to substitute for informal sources (Macoloo 1994, Struyk and Turner 1986). However, the higher costs of formal projects present households with a funding gap that contributes to the survival and even flourishing informal sector (Baken and Smets this volume, Bond 1990, Macoloo 1994).

SELF-HELP HOUSING AND INFORMAL FINANCE

The virtues as well as the shortcomings of self-help housing have received considerable attention in the literature. This attention has concentrated upon *inter alia* the effectiveness of self-help as a means of providing adequate shelter to low-income households, its relationship to the development of capitalism and to the abrogation of state responsibility, and its implications for gender relations and for household formation. In general, these issues and debates do not need to be rehearsed here. Rather, we wish to concentrate on only those issues that relate to the provision and management of housing finance.

One of the key observations made of self-help housing is that it allows households to build in stages in order to 'synchronize investment in buildings and community facilities with the rhythm of social and economic change' (Turner 1967: 167, 1976). Self-help, therefore, was regarded as housing that was 'affordable', relative to the size and stability of existing income over the short term and the household life-cycle. Although advocates were careful not to claim that self-help was synonymous with 'cheap' housing, it was argued that the incremental building process allowed costs to be kept as low as possible. An often-repeated claim was that self-help housing cost approximately one-half of contracted housing (Turner 1967). Turner argued that the elevated costs of contracted housing were due to unrealistic standards and bureaucracy, compared to the non-remunerative 'sweat equity' of self-help (Turner 1976, 1982, Ward and Macoloo 1992). Later work, however, noted that most self-help is actually constructed with the paid assistance of artisans and, for specialist tasks, skilled labour. Moreover, according to Burgess (1982: 65, 1985), the reduced costs were partly attributable to households not having to pay interest on finance to property and industrial capital. In fact, neither advocates nor critics of self-help have provided rigorous data on the costs of housing or finance. It is unclear, for example, whether Turner's data include interest paid to informal financial sources, some used for only a few days, over the twenty or more years of self-help construction. Nor, in making the comparison with contracted housing, is it clear how subsidies were dealt with or whose data were used when many public housing institutions appear not to have much idea of the real cost of the units they supply.

The advocates of self-help did not argue for house construction to take place without external finance. Indeed, in his early analysis of the housing problem Abrams (1966) identified the lack of finance (specifically low savings and high interest rates) as major obstacles to progress in housing provision. Turner (1982: 101) also cited one of the lasting achievements of a project in Peru as the establishment of a local housing financing system. Yet, elsewhere the attitude toward finance appears to be sceptical, especially with regard to households using finance other than collective small savings (Turner 1976, Turner and Fichter 1972). Indeed, Turner argued that house construction with the aid of a mortgage may reduce a household's security of tenure and that debt payments could undermine the ability of households to reduce their outgoings at times of crisis, an important principle of self-help (Turner 1967: 11). Therefore, advocates of self-help were keen to encourage efforts to save as a signal to potential lenders of creditworthiness, but urged care about how this transition was to be managed (UNCHS 1991a).

The self-help literature presents a symbiotic relationship between finance and housing consolidation. The argument is that investment in housing will confirm a household's security of tenure which will have a positive psychological influence upon saving and result in further improvement (Abrams 1966, Turner 1967, 1982). Yet, this simple notion of housing consolidation highlights a problem of finance. In order to confirm security through possession households are obliged to build a rudimentary construction soon after occupation, in effect substituting affordability for quality in the short term. But, this initial investment may take a substantial proportion of surplus income and/or savings and may imply significant opportunity costs, especially as the initial construction is soon replaced (Gough this volume). As Turner (1968: 357) noted, there is also the likelihood that households will be unable to recover the full value of their investment unless the property is legally recognized. Households, therefore, may be faced with expensive housing that cannot be used as collateral on a mortgage for further improvement, or be passed on as an inheritance or resold to allow residential mobility (Gilbert forthcoming). By contrast, the advantage of even a small quantity of finance is that it allows households to front-load the construction of a high-quality construction (Pugh 1994, Ward 1981).

Despite the attention afforded to housing finance by early writers on self-help, this seems to have been downplayed in later literature. Instead, the emphasis moved into the technical design of housing and materials, strategies for community organization and service provision, and theoretical concerns for commercialization (where finance received limited attention) (but see Burgess 1985: 291–3, Ward and Macoloo 1992). By the early 1980s, one could argue that we knew more about how self-help housing was built, and almost as much about its role as a catalyst for political consciousness-raising, than about how it was paid for. The low priority given to housing finance probably reflects the tendency for self-help debates to be engaged by architects, planners, geographers and sociologists rather than economists. Many studies sought to identify and weigh the variables considered to be most significant for the rate of house consolidation, affording considerable attention to demographic (age, household structure, education), settlement (tenure, legality, existing service provision) and land market characteristics. Logically, strong links were found between consolidation and income, but few studies systematically concerned finance. Yet, when finance was the focus of the research, it was found that a lack of finance was twice as important to house improvement as security of tenure and was the most important constraint mentioned by those households most likely to improve their housing (Wegelin and Chanond 1983).

Although we would argue that housing finance was under-represented in research, it was a critical component of policy especially in sites-and-service and upgrading programmes. In order to ensure that the eventual housing was affordable to low-income households, most of these programmes involved substantial subsidies through the provision of land, building materials, technical assistance and repayment holidays (Datta this volume, Pugh 1994, Rodell 1983). While it was envisaged that these 'hidden costs' would be recouped from taxation, this was rarely achieved (Baken and van der Linden 1993, Devas 1983). Indeed, during the lifetime of many projects it was frequently the raft of financial instruments devised to retain affordability and motivate house construction that presented institutions with the greatest administrative and political headaches (Bamberger and Harth-Deneke 1984,

Macoloo 1989, UNCHS 1991b). In part, these difficulties may have been the result of a preference for complicated financing measures such as cross-subsidization rather than simpler instruments that provided households with an income or opportunity to capture windfall gains to pay project costs (Ward 1981). For households, however, the instruments proved insufficient to prevent the need to resort to informal sources, to cut back on consumption, take extra employment or sell up (Macoloo 1994, Ward 1981).

Housing finance: a problem of savings?

The role of savings in the acquisition of housing by low-income households has been recognized in a variety of studies (Merrett and Russell 1994, Struyk, Katsura and Mark 1989). Macoloo (1994: 290) reports that two-thirds of households in Kenya used savings to purchase construction materials, making this the most important source of finance during the initial stages of house-building. Yet, the significance of savings to the housing process is not without its problems. As other studies show, many households are able to become 'owners' early on, before they would be able to develop a savings profile. While it is increasingly accepted that low-income households are not too poor to save, in an environment of low and insecure incomes, and rising prices for building materials, land and services and rising taxes, it seems unlikely that households would amass sufficient savings on a consistent basis to acquire or build housing.

This would seem to be supported by macro studies that show low propensities to save in the poorest countries and in the poorest households in all countries (Edwards 1995). Most low-income households lack safe and convenient methods to save and many institutions insist upon minimum balances or do not offer positive returns on savings. Indeed, in some circumstances low-income households are charged to deposit savings, either directly or through negative interest rates (Johnson and Rogaly 1997: 19–20). In order to save in the formal system, therefore, low-income households have to possess a minimum level of funds and an income sufficient to pay charges: in the short term saving may be a net drain on household resources.

The lack of institutional capacity is not the only reason for the low rate of saving. Countries such as Chile, Singapore and Malaysia have implemented sophisticated financial products to increase the ratio of money to GDP and raise the rate of saving (Edwards 1995). Such reforms, however, have not improved conditions for those on the lowest incomes as measures have not been taken to improve the unequal distribution of income: with no surplus income, savings will be inelastic in relation to interest rate changes. Furthermore, anecdotal evidence suggests that, in some countries, many low-income households are heavily in debt, and are therefore unable to be net savers, and have learned through experience to be highly distrustful of financial institutions (Siembieda and López this volume).

The difficulty experienced by households trying to save in the context of limited institutional capacity suggests that many must be holding savings outside of the formal financial system (Boleat 1987: 159, Okpala 1994, World Bank 1993a). There is very little research, however, on the form in which these savings are held although consumer items and jewellery as well as cash are mentioned frequently. One further possibility is that households invest in housing as a surrogate form of saving believing this to be a reliable store of value (Gilbert forthcoming). As consolidation takes place the belief is that most properties will appreciate in value over time, although the small size of the second-hand property market makes reliable assessments of value difficult and some suspect that the real trend may be quite flat over the short term (Gilbert forthcoming, Gough 1998). Thus, while there is evidence for a link between financial depth and housing investment, at the micro level, it is the lack of financial capacity that may increase investment in housing. We are, therefore, presented with a paradox that households need to save in order to build adequate housing, but that some appear to be over-investing in housing as a form of savings.[12]

There is, of course, the possibility that researchers have interpreted 'savings' too widely. Households are often reluctant to discuss domestic finance. The stigma of debt, especially where this involves moneylenders, may motivate households to refer to savings as a euphemism for a host of domestic resources. In cases where money has come from kin, it may be particularly difficult to distinguish between a loan and a gift; while the latter might be considered to be a genuine saving, the former should be a liability even though repayment might be waived (Baken and Smets this volume). Unless researchers have made the distinction clear, it is advisable not to draw too hard a line between savings and informal finance as a means to pay for housing.

Informal finance

Informal savings and loan associations are common to many developed and developing countries and are known by a variety of names: *chilimba* in Zambia; *gamaiyah* in Egypt; *hui* in Vietnam; *kusukus* in Ethiopia; *paluwagon* in the Philippines; *susu* in Ghana; and *tanda* in Mexico (Bouman 1995, Johnson and Rogaly 1997, Merrett and Russell 1994, O'Reilly 1996). In the past, the presence of informal finance was regarded as an intermediate stage which would be replaced by formal finance once development was under way. Today, however, informal finance is increasingly looked towards as a potential means of solving a host of development problems (Adams and Fitchett 1992, Bouman 1995, Rogaly 1996). Neo-liberals admire its self-regulatory nature, entrepreneurial ethos and pragmatism, while advocates for broader social development agendas support informal finance for providing a vital security network for low-income households who have to rely upon local rather than bureaucratic networks and knowledge (Ardener and Burman 1995, Copestake 1996).

Although there are considerable variations concerning size, charges, methods of access, propensity for fraud and permanency, most informal systems operate by regular cash contributions into a common pot which can either be called upon in the event of an emergency or be allocated as a lump sum according to a prearranged cycle with the last disbursement ending the group. These 'rotating' systems, or rotating savings and credit associations (ROSCAs), allow new members to substitute for members who no longer need lump sums or who are unable to guarantee regular contributions. In the more sophisticated ROSCAs disbursements are made as loans. Repeated short-term loans enable borrowers to adjust sums to changes in household income, thus keeping debt/income ratios to within manageable proportions – parallels with the key principles that underpin self-help, namely, dweller-control, scale and flexibility are immediately apparent. There is also the possibility of 'loan ratcheting', the promise of larger loans as an incentive to complete initial smaller ones (Merrett and Russell 1994). Where ROSCAs hold substantial capital sums (estimates vary up to one-third of all savings) and external support is available from government agencies or NGOs, special investment instruments have been established to raise additional capital on the formal financial markets (Jones and Mitlin this volume, UNCHS 1991b).

Such innovative forms of group lending can present a number of important advantages over institution-based systems. Access to finance can be dictated by group membership, thus offering opportunities to exclude better-off households or those likely to default, and social pressure to repay loans can be exerted through regular meetings while group organization can reduce administrative costs (Bouman 1995, Johnson and Rogaly 1997). Finance obtained from ROSCAs is rarely prescriptive but can be diverted in a flexible manner to a variety of purposes including the purchase of consumer goods, crisis needs such as funeral costs, income generation and payment of existing debts (Ardener and Burman 1995, Schrieder and Cuevas 1992, Titus 1997). In relation to housing, research shows the importance of informal finance (Merrett and Russell 1994, UNCHS 1991a). In their study of Jordan, Struyk, Katsura and Mark (1989) found that formal and informal finance (including intra-family loans) accounted for about one-half of finance sources. In Kenya, Macoloo (1994) found that informal finance was a critical component of household budget strategies even for settlements where formal finance programmes had been introduced. There is also substantial evidence of a link between access to informal finance, income generation and house improvement (Albee and Gamage, Cosgrove this volume).

This research notwithstanding, one must be wary of romanticizing the role and potential of informal finance. As Okpala (1994: 1578) has noted, the enormous housing shortage is evidence that, despite its significant contribution, informal housing finance is unable to cope with the task. Moreover, there are distinct social costs associated with informal finance. Group lending operates through a process of group or self-exclusion that denies membership to the very poorest households for fear of debt or default. ROSCA-style programmes are also vulnerable to the effects of inflation as deposits are made in cash. As observed by one group of experts, 'setting up "revolving funds" is a very fashionable developmental creation but many of these funds do not actually "revolve"' (HIC/ACHR 1994: 16, also Schmidt and Zeitinger 1996a: 252). This loss of real value has consequences for the recipients of finance who are able to buy fewer materials as prices rise over the ROSCA term.

An important feature of group finance schemes is the dominance of women. Many of the most innovative schemes are run by or for women and some observers ascribe the success of these schemes directly to observations that women are more astute fund managers, display greater probity where small sums of money are concerned and have a greater awareness of day-to-day problems (O'Reilly 1996, Schrieder and Cuevas 1992, Titus 1997).

GENDER AND SELF-HELP HOUSING

A particular feature of housing and finance research has been the lack of attention to gender and an avoidance of the need to deconstruct the social relations of housing (Moser 1992a, Tinker 1992). The literature on self-help, for example, conceptualized low-income communities and households in homogenous terms, implicitly downplaying the important role of women in the housing process. This has had a number of consequences. First, the concern for a growing commodification of housing did not recognize that this would affect women especially due to their relative lack of construction skiils (and so reliance upon hired labour) and fewer social networks outside the settlement (Miraftab 1994). Second, urban planning and housing programmes have tended to be based upon stereotypes of the household, ignoring the needs of women, who suffer disproportionately from a lack of information, less familiarity with complicated application procedures, discrimination on the part of bureaucrats and inequitable institutional arrangements (Moser 1992b, Young 1993).[13] Third, so-called 'practical' gender needs were afforded less attention as research had not revealed the triple role that women perform as reproducers, producers and community managers, which creates distinct housing interests and decisions from men (Miraftab 1994, Todes and Walker 1992). As reproducers, women spend more time in and around the house, so its condition is of great concern to them (Chant 1997, Tinker 1992). Similarly, as producers and community managers, settlement layout, service provision and the lack of child-care provision particularly affect women (Moser 1992b).

Researchers also failed to recognize the apparently strong link between gender, poverty and housing that should have alerted them to the development implications of policy decisions (Appleton 1996, Smillie 1995). While research identified an 'urbanization of poverty' it was much slower to comprehend a 'feminization of poverty', despite figures showing that between

17

1965-70 and 1988 there was a 47 per cent increase in women living below the poverty line as compared to a 30 per cent increase in men (cited in Jackson 1996).[14] The greater vulnerability of women to poverty, their inferior labour market position and difficulties of maintaining welfare levels have drawn attention to the heterogeneity of household structures (Chant 1997). In particular, research has concentrated upon the female-headed household which some estimates put at one-third to one-half of households in urban Africa and Latin America (Buvinic *et al.* 1983, Chant 1997, Tinker 1992). Contrary to conventional perception, research indicated that household type was not an automatic indicator of welfare as resources are allocated more effectively and dependency ratios were sometimes lower (Appleton 1996, Chant 1997, González de la Rocha 1994).

In turn, the attention to female-headed households has provoked criticism for concentrating on one household type – the young, single, mother – and ignoring women living with their adult children or those living in nuclear households (Varley 1995, 1996). Another group to have received little attention is elderly women, many of whom live alone. Increased life expectancy has created significant elderly populations which, due to the non-existence or collapse of social security systems, and limited savings due to employment in poorly paid or non-remunerative jobs, are among the poorest urban residents. While some researchers have argued that the inclusion of elderly relatives into extended households has had positive outcomes by taking child-minding duties away from younger members, allowing them to work, dependency ratios have not fallen dramatically and per capita income levels remain among the lowest (González de la Rocha 1994). As is shown for Singapore, the elderly, who include disproportionate numbers of women, present a unique set of housing and finance problems (Addae-Dapaah this volume).

Finally, the tendency to concentrate on the household has created an illusion of an equal distribution of power and resources within the household (Dwyer and Bruce 1988, Moser 1992b). However, as the hierarchical nature of domestic politics has been uncovered, the distribution of power and resources has been found to be highly unequal (albeit often accepted by women due to a (false) perception of their limited contribution to household welfare) (Kabeer 1991, Sen and Grown 1987). The addition of resources through housing finance programmes must not, therefore, be considered neutral in relation to domestic power conflicts and decision-making. The impacts of programmes of housing construction or improvement will be mediated by the ability of household members to share information, prioritize certain resource allocations and manage time and money effectively. Under certain circumstances the addition of finance will have dramatic effects upon the ability of women to control resources and decision-making – possibly leading to their empowerment.

Gender, finance and empowerment

'Empowerment' is one of the most frequently used terms in the development lexicon and is widely, indeed rather nebulously, used to signify the greater participation and political equality of women (Ackerly 1995, Kabeer 1994, Smillie 1995). A number of researchers have identified finance as an important and effective instrument to raise levels of empowerment by giving a greater stake and voice in the household decision-making process and autonomy from the needs of others. Yet, some advise caution and a degree of scepticism has been expressed as regards the direct link between finance and empowerment. As noted by Ackerly:

> 'Empowered', the borrower wisely invests money in a successful enterprise, her husband stops beating her, she sends her children to school, she improves the health and nutrition of her family and she participates in major family decisions.
>
> (1995: 56)

Even assuming such a finance/empowerment relationship, it is extremely difficult for researchers to gain either quantitative or qualitative indications of the extent or direction of change. Most researchers are therefore obliged to resort to surrogate variables, principally changes to income or health (Dwyer and Bruce 1988, Smillie 1995). As Goetz and Gupta (1996: 45) argue, 'women's continued high demand for loans and their manifestly high propensity to repay is taken as a proxy indicator for control and empowerment.' The problem is the circularity of an argument that a continued desire for finance is an indicator of success, rather than a sign of the conditions of poverty and powerlessness that force women to demand finance in the first place.

Arguments have been made, therefore, to distinguish between economic and social empowerment. It is unclear, for example, whether finance is an appropriate or desirable means to challenge gender roles as, by definition, debt is a liability and may serve to integrate women into patriarchal economic systems (Rogaly 1996, Sparr 1994). The ability of women to retain control of the finance, therefore, is critical. Research shows that for the extension of finance to women to translate into more equitable gender relations within the household is dependent upon a woman's existing status and self-confidence (Ackerly 1995, Kabeer 1994). Without this status and control, studies reveal the danger that women's resources disappear into 'household' budgets to be used by a husband, partner, son or mother-in-law (Ebdon 1995, Goetz and Gupta 1996). Moreover, despite losing control of the resources, many women continue to make their contributions even at the expense of personal hardship, occasionally resorting to moneylenders, pawning assets and engaging in malpractice (Ardener and Burman 1995, O'Reilly 1996).[15] One alternative that provides an opportunity for women to maximize control and to build self-esteem, as well as put pressure on male-dominated institutions to become more gender sensitive, is the formation of women's-based organizations usually in association with NGOs (Tinker 1992, Todes and Walker 1992, Pohlmann 1995).

NGOs AND HOUSING FINANCE

Until the 1980s, the majority of NGOs steered clear of housing and finance issues. For most, the provision of housing was too complex, both technically and politically, requiring the acquisition of land, construction materials and labour in the short term and services in the longer term. An understanding of planning procedures and regulations, and involvement in the unfamiliar terrain of municipal politics and business practices, were unattractive propositions to staff that had little practical training in economics, law or administration (Abugre 1993, Johnson and Rogaly 1997). The requisite long-term commitment to housing finance was also contrary to the ethos of many NGOs that believed in short-term intervention as a sign of success.

Nevertheless, NGOs have become increasingly involved in housing finance either as part of lending and savings programmes for micro-enterprises or as an important element in broader poverty-reduction strategies (Jones and Mitlin this volume). Without wishing to doubt the achievements of these programmes, there are some important gaps in our knowledge about NGO performance. One of the most serious concerns the lack of systematically collected and standardized information that can be subjected to independent scrutiny. Instead, the NGO case is often presented using subjective appraisals through brochures, videos and other media (Bouman and Hospes 1994: 13, Dichter 1996). As a result, some observers feel that there has been a tendency for advocates to 'have encouraged more self-congratulation, or even romanticism, than realism about NGO financial management capabilities' (Abugre 1993: 159). We need more definitive evidence of performance and effectiveness, and a clearer balance-sheet of best practice.

A particular note of concern relates to the repayment rate as a meaningful indicator of institutional capacity and as a proxy for programme sustainability (Havers 1996). NGOs often employ vague terms such as 'overdue' or 'outstanding', inconsistent criteria such as the threshold for converting 'arrears' into 'default', and average data that make no mention of membership turnover (Havers 1996, Hulme and Mosley 1996). On closer inspection, the experience of some micro-finance programmes indicates high repayment rates but poor overall performance (Hulme and Mosley 1996, Schmidt and Zeitinger 1996a, von Pischke 1996). It would seem that the experience of many housing finance programmes is very similar. The repayment rate of a housing finance programme in South Africa, for example, was estimated to be 125 per cent due to high levels of prepayment by some households that obscured delayed payments by 10-15 per cent of members (Mitlin 1996: 7–8). Even taken at face value, there has been limited research conducted on the reasons for high repayment rates.[16] As Goetz and Gupta (1996: 54) have warned, in the case of women borrowers, high rates may have less to do with profitable loan use than with a desire to retain membership of one of the few social and public institutions to which they have legitimate access beyond the household.

A critical subject upon which more information needs to be made available and assessed is the explicit and implicit subsidies in NGO programmes. From the few comparisons made to date, it seems likely that subsidy levels for NGO programmes are low when compared to government housing programmes and can be allocated more efficiently than if external agencies tried to select recipients (HIC/ACHR 1994, Mitlin 1996). But, there are cases of NGOs that charge no interest on loans, which charge flat rates under inflationary conditions or which appear to make no detailed assessment of administration costs even when it is clear that these are high (especially as a proportion of revenue) and are not being passed on to borrowers.[17] While many NGOs do not hide the fact that they operate with substantial levels of subsidy and are able to justify this by their concern for social development, this should only serve to encourage transparency.

Of course, it is very difficult to use standardized measures in order to assess the qualitative services that form an important component of NGOs' 'comparative advantage' (Edwards and Hulme 1992a). Compared to other institutions, NGOs have been shown to design small-scale programmes that closely match the needs of local communities and to manage these in ways that retain the flexibility to innovate according to changing circumstances. Part of this success is attributed to the effectiveness of NGOs in creating or strengthening community-based programmes despite lower levels of unity exhibited by urban households according to kinship or ethnicity, or where there has been a history of bond- or service-payment strikes (Lee 1995, Vakil 1996). NGOs claim that community programmes can engender a sense of stake-holding, reduce transaction costs per loan (a particular problem when small quantities of finance are being allocated) and raise gearing levels as household savings can be pooled to convince financial institutions to lend additional sums. Case studies that employ qualitative research methodologies would serve to strengthen these claims.

Such studies might consider putting less emphasis on programmes identified in advance as 'successes' and look in addition at cases of failed or failing programmes. Analysis of the Group Credit Company in South Africa, for example, reveals the reasons why a Grameen-clone faced collapse after only a few years of operation in urban settlements. Although successful with the first tranche of small loans, the company found that the groups quickly disintegrated as members had little in common other than finance and as mobility levels were high. Ultimately, the company was forced to abandon group lending and to make funds available to individual borrowers able to contribute a much higher proportion of their own funds. Other studies might be able to identify at what point the advantages of 'scaling up' start to diminish and programmes are subject to the growing 'errors of excess' (Berger 1989: 1026). Scale can make NGOs susceptible to ambitious programmes beyond their means, relax their loan criteria, and threaten the sense of 'stake-holding' and participation of low-income groups, and the desire of NGO staff to innovate (Bouman 1995: 374–5, Ebdon 1995, Rogaly 1996, Tendler 1989).

We also need greater insight into the impacts of existing programmes on housing construction. While NGOs are able to demonstrate many achievements, few have managed to provide large numbers of housing loans. Of twelve important NGO-sponsored housing finance programmes, the maximum number of households assisted by any single programme was 2,300 and many of even the largest independent NGO programmes manage to assist fewer than one hundred households in any year (HIC/ACHR 1994). As already mentioned, most studies have been unclear and inconsistent on the outputs from finance initiatives, preferring to present data on the number of housing loans rather than houses (many loans can go to the same household) or distinguishing between house improvements and new construction.

Finally, more concrete indications need to be provided of the association between finance, housing and empowerment through NGO programmes. For this to happen, researchers might consider the following points. First, whether NGO concern for empowerment can be squared with the suspicion that they are attracted to supporting women's organizations because of the presumed passivity of women (Ardener and Burman 1995, Johnson and Rogaly 1997: 38). Second, whether as actors with agendas set by national or international donors, the act of NGOs *bringing* empowerment to marginalized communities is a contradiction in terms (Kabeer 1994). Third, whether NGOs need to work within existing gender structures until programmes are established before affording priority to empowerment.[18] For example, the work of NGOs such as the Grameen Bank have tended to conform to established gender roles by speaking to the family of prospective women members and stressing to them the benefits that participation will bring to the households. Fourth, whether under certain conditions the provision of finance disempowers women, for example by adding a fourth burden of financial management, requiring attendance rates at meetings at considerable cost in terms of time, lost income or domestic dispute. Fifth, whether gender research has been essentialist, treating all women the same and ignoring evidence that women are capable of being exploitative of each other. As Hulme and Mosley (1996) argue, the provision of finance is likely to empower some women *vis-à-vis* other women, but not compared to men.

CONCLUSION AND (FURTHER) INTRODUCTION

Housing finance has moved to the top of the international urban policy and research agendas. Following the poor performance of formal institutions, reforms appear to be delivering greater innovation and greater financial robustness in both the public and private sectors. NGOs too have found greater institutional and political space, and have established programmes that adapt informal finance methods, promote savings among low-income households and can achieve sustainability over the longer term. Finally, researchers are more attuned to the gender implications of policy and the potential for finance to change domestic decision-making, improve resource allocation and reduce poverty. Housing and finance appear to offer a useful combination of mechanisms that can enhance empowerment.

To raise the profile of housing finance still further will require greater attention to theory. The literature has developed theoretical insights from work conducted on micro-finance (Hulme and Mosley 1996) and the economics of formal housing finance (Buckley 1996, Buckley and Mayo 1989). Strangely, there seems to be no work that has used regulation theory and little that has borrowed from new institutional economics. What, for example, are the transaction costs inhibiting private-sector involvement, what are the limitations imposed by legal frameworks, and how have political factors intervened to raise risk? In addition, there is little work that links housing finance to social theory even though group lending raises issues of power, trust and collectivity, to institutional development and grassroots action. The lack of theoretical framework may well result in less perceptive policy recommendations in the future (Ward and Macoloo 1992).[19]

Although we are unable to substantiate this claim, the lack of suitable theoretical frameworks might be contributing to problems of empirical rigour. In drawing direct causal link between finance and housing, for example, researchers have not looked at how finance impacts upon other financial flows within the household. Research, therefore, has not managed to isolate the loan-effect from extra incomes, reduced expenditures, better overall management, inheritance, remittances, windfalls or other loans. It is equally difficult to know what might have happened had loans not been made available or had a different type of loan been offered.

Finally, we need more awareness of intra-household negotiations over finance decision-making and loan use. The addition of finance is likely to change domestic decision-making substantially by reordering perceived priorities and changing power structures. Yet, we know very little about how these changes affect other household members and whether households become more competitive or co-operative as a result. A subject upon which there would appear to be almost no explicit research is the role of men, especially in households where the woman has received finance. This would seem to be storing up problems for the future. We know that some institutions are reluctant to lend to low-income households due to the perceived financial recklessness of men and that others have created women-only programmes either to aid sustainability or to enhance empowerment. But, as Goetz and Gupta have commented:

This instrumental approach to women as conduits for credit for the family plays on, and reinforces, traditional cultural notions of womanhood, with women seen as moral guardians of the household and policers of recalcitrant men. The implications of this process for the way men are being constructed culturally in relation to credit operations are also disturbing.

(1996: 55)

In the longer term, it would be useful to know why men act in ways that are against household (and possibly their own) interests and how best to raise general levels of participation and co-operation while allowing women to control a greater share of resources and decisions.

NOTES

1 For discussion of the principles and performance of these reforms see Baken and van der Linden (1993), Malpezzi (1994), Jones (1996), Jones and Ward (1994) and World Bank (1993b).

2 There are considerable problems with the methodology used by the Housing Indicators Program: for example, attempting to globalize unweighted proxy statistics based on consultant estimates (note the low figure for squatter housing). The data also fail to indicate normative conditions, showing only that measures for one country may be 'high' or 'low' relative to others, and provide no indication of trends which are probably more important to policy-makers.

3 There are problems with how one measures household income (including from the informal sector, in kind and subsidies), what is meant by the 'household' in cases where multiple, related and unrelated, people occupy a 'family' compound or there are multiple decision-makers and many conflicting interests (Dwyer and Bruce 1988, Varley 1994).

4 In the extreme case of eastern Europe, in 1990, eight of the eleven transitional economies had only one bank prepared to make housing loans (USAID 1997).

5 Although research has concentrated upon housing institutions or banks, insurance companies, building (friendly) societies and the Post Office also lend considerable sums (Rakodi 1995, UNCHS 1991a).

6 The use of contract (payroll) funds was adopted in developed countries to channel resources towards or away from dedicated sectors of the economy and because short-term deposits would be insufficient to fund long-term debt (Diamond and Lea 1992). In developing countries these conditions were even more acute especially given the demands for finance from industry and low savings ratios. The decision not to index payroll contributions was a useful device to avoid the pressure from employees for the expected housing to be delivered.

7 Chronic inefficiency is demonstrated by Ecuador's Housing Bank, which delivered only two mortgages per employee throughout the 1980s (Klak 1993).

8 Some notable examples of low assets per capita are Bolivia (US$ 8), Kenya (2.95), the Philippines (5.32) and Korea (8.96). Among the few countries to record increased assets are Uruguay (from US$ 93 to 492.6, 1985–9), Jordan (130.94 to 305.32, 1979–89), Malaysia (17.3 to 44.5, 1979–85), Singapore (476.39 to 6,490.77, 1979–85) and Thailand (2.26 to 18.22, 1979–89).

9 There has been a high degree of convergence in policy advice given to developing countries with strategies adopted in Europe and the USA. The deregulation of financial

markets in developed countries has led to greater elasticity in the supply of funds, real-cost pricing and a wider range of funding instruments. Reform, however, has also produced greater interest rate volatility, enhanced a tendency toward over-indebtedness and accentuated property price fluctuations (Ball 1990, Diamond and Lea 1995, Hamnett 1994). Although Diamond and Lea (1992) argue that these effects are likely to lessen as the transition toward market mechanisms proceeds, the fact they have been so acute does not bode well for the more segmented and imperfect financial markets of developing countries.

10 Recent events in Mexico and Thailand have demonstrated that the threat to the economy of financial mismanagement is not restricted to the public sector. In Thailand, the finance and property sectors have been vulnerable to the effects of maintaining an over-valued currency, declining economic growth rates and conflicts between regulatory authorities. Although the link between financial institutions and property developers has been one of the successes of Thailand's urban development, poor management has produced an unstable finance-led property boom that has resulted in 20 per cent of loans being classified as 'non-performing'.

11 We know very little about the size or impact of non-finance subsidies that are especially prevalent in urban settings: namely, free or cheap land, building materials or services.

12 As long ago as 1966 Abrams observed that low-income households tended to spend money on goods rather than save in the belief that goods held value relative to inflation.

13 An important issue is legal status. In some countries, legal rights to property are granted exclusively to men and on death to their male children (Miraftab 1994). In general, women are much less informed about their rights than men and especially where there is a significant gap between legal rights and the cultural acceptance of women to own property (Miraftab 1994, Tinker 1992).

14 Jackson (1996) actually disputes the 'feminization of poverty' thesis, the strength of which depends upon the criteria used to define the head of household. Thus, *de facto* female-headed households who receive remittances from absent men may be better off than male-headed households.

15 According to Ebdon (1995: 53) the high repayment rates among women may be due to the earnings from husbands who use the loans and have little to do with the gender of the borrower.

16 The literature provides few indications of how NGOs distinguish good from bad borrowers. The reliance of many NGOs on personal contact and tacit knowledge appears to be justified, but the experience of financial institutions in developed countries is that these are poor predicators of credit-worthiness and need to be backed up with 'pooled equilibrium' strategies.

17 Although the data are not presented in such a way as to make a conclusive statement, it would appear that 25 per cent of the interest on loans from a Colombian NGO was required to cover costs and that a Brazilian NGO was carrying administrative costs equal to 35 per cent of its portfolio (HIC/ACHR 1994).

18 While advocates for a minimalist position argue for programmes to limit their remit to finance-only services, the most successful micro-finance programmes have tended to originate from this position and have subsequently added extra services (Tendler 1989).

19 This lack of concern for theory in the housing finance literature follows a well-established pattern in the housing literature of downplaying 'pure' research and placing policy-driven work top of the priority list (Ward and Macoloo 1992).

2

HOUSING FINANCE AND NON-GOVERNMENTAL ORGANIZATIONS IN DEVELOPING COUNTRIES

Gareth A. Jones and Diana Mitlin
University of Wales, Swansea and the International Institute for
Environment and Development, London

INTRODUCTION

Although housing conditions represent one of the most visible signs of poverty in developing countries, formal financial institutions have been reluctant to offer their services and official development organizations afford urban issues a low priority.[1] Seeking to address this gap, during the last twenty years there has been considerable innovation in the provision of housing finance to low-income households, particularly among Southern NGOs working in urban areas (Anzorena 1993, Arrossi *et al.* 1994). NGOs that have been involved in the implementation of upgrading and service delivery programmes as well as more radical urban development strategies involving community mobilization and empowerment have extended their activities into housing finance programmes (Cabannes 1997, Cruz 1994, Dizon 1997).[2]

The development of housing finance programmes by NGOs has taken place within a broader context of NGO initiatives in micro-enterprise credit and loan finance for small and medium-sized entrepreneurs (Adams and Fitchett 1992, Schmidt and Zeitinger 1996a). These programmes have received considerable attention from donors, so that the sector had become sufficiently significant to hold the World Micro-Enterprise Conference in 1987 and the World Microcredit Summit in 1997. A number of these NGOs and community organizations with an established track record in micro-finance have diversified towards housing finance, motivated in part by the realization that their borrowers were investing loans given for income-generation into housing and basic services (often in contravention of the rules). For example, by 1995, the Grameen Bank in Bangladesh had provided 330,000 housing loans to the members of its savings schemes; in India the Self-Employed Women's

Association (SEWA) now also provides housing loans and Banco Solidario (BancoSol) in Bolivia has identified housing as one of its intended areas for market expansion (Anzorena 1996, Mutua *et al.* 1996).

The diversity of the initial engagement with housing finance is repeated in the methods of operation with some NGOs adopting community-based approaches, others the promotion of savings mobilization or credit rotation, direct institution-beneficiary models or the provision of technical consultancy and intermediation (Mitlin 1996). However, one can observe that many housing finance programmes share a number of common characteristics. In general, NGO loan programmes are one part of more substantive programmes to improve housing conditions that may involve the provision of technical assistance, community development training, grants for improving infrastructure and services, building materials production and support in negotiations with local authorities. In some cases, NGOs have combined housing finance with the availability of emergency loans and finance for micro-enterprise development (Albee and Gamage, Cosgrove, Gordon this volume).

At some risk of over-generalization, other similarities also stand out. Before providing housing loans, NGOs typically request borrowers to demonstrate a propensity to save, offer a financial subsidy with the loan funds, either through interest rates cushions or grants, and offer loans for a range of activities related to housing development. In some cases, only part of the housing development process is supported through loan finance with the household being expected to raise the remaining monies from other sources, or several loans may be provided in stages as land is purchased and building material costs are repaid (Albee and Gamage 1996, Cabannes 1997).[3] The reasons for the provision of housing finance are as diverse as those for the provision of micro-finance. Some organizations wish to improve living conditions, others to secure empowerment, others to extend financial services.

This chapter looks at the experience of NGOs working with housing finance, either as the direct providers or as intermediaries, as well as those NGO-style approaches that have been taken up by governments through the provision of collective loans and the strengthening of community-based organizations. These experiences are discussed within a broad framework of micro-finance initiatives and poverty-reduction strategies. The chapter summarizes recent trends in micro-finance and explores the relationship between NGO programmes for income generation and housing finance. It considers why and how housing finance has been used within poverty-reduction programmes. We pay particular attention to the methods used by NGOs to scale up their activities. In contrast to many micro-finance initiatives for income generation which have sought to scale up through the use of private finance and market strategies, NGOs working with loan finance for housing have tended to seek government support to achieve more widespread coverage. In several countries this strategy has been successful with major government programmes drawing on NGO models.

Our general aim is to provide an assessment about the potential for NGO-backed programmes to deliver affordable and sustainable housing finance. While there is considerable advocacy for NGO participation in housing finance at the present time, we need to be more certain of the circumstances in which it is appropriate for NGOs to intervene and in what form they should do so to achieve prescribed goals. Therefore, while accepting that there is a natural institutional tendency to transfer perceived 'successful' practice, a critical look at the experience of micro-finance identifies a number of problems with existing programmes that suggest a need for caution before advocating particular role-models for housing finance programmes.

MICRO-FINANCE: TRENDS AND PERSPECTIVES

The past few years have witnessed a growing interest in the ability of micro-finance programmes to deliver small quantities of capital (sometimes as little as US$ 10) at short notice to low-income households in ways that have the potential to be sustainable over the long term. Once micro-finance programmes are in operation, assessment of their effects at the household level points towards the more equal allocation of income, a reduction of the costs associated with cyclical or unexpected crises, and assistance with short-term consumption needs (Bennett and Cuevas 1996, Hulme and Mosley 1996). In addition, studies reveal that the delivery of micro-finance loans may provide further benefits through improved support for enterprise formation and growth, with subsequent benefits for low-income men and women as both employees and consumers (Gibson 1993).

Many of these programmes are based upon development strategies that seek to improve the efficiency and effectiveness of financial markets. Advocates argue that profitable investment opportunities are forgone due to the reluctance of formal financial institutions to lend to low-income groups (Remenyi 1991, ESCAP 1991). Frequently cited reasons for this institutional reluctance range from, *inter alia*, a cultural resistance to accepting that low-income groups are credit worthy, a belief that small savings and loans will mean high transaction costs, and the lack of acceptable forms of collateral (Rhyne and Otero 1992). With this understanding, NGOs have developed financial institutions designed to ensure the better functioning of financial markets for those on low incomes. Such financial institutions seek to correct directly the deficiencies of the formal sector by trying to ensure that bank branches are located where they are needed, are open at convenient hours and staffed by people trained to work with their potential clients. On the finance side, micro-finance institutions accept more flexible forms of collateral and in many cases encourage small group formation to enable collective and personal guarantees. These approaches are supported by research that demonstrates that households have a strong desire to save, that transaction costs can be kept to a minimum and collateral can have a 'social' instead of a purely financial base (Adams and Fitchett 1992, Remenyi 1991).

The ascendancy of 'minimalist' micro-finance

While the aforementioned characteristics are common to most micro-finance programmes, Copestake (1996) has distinguished three perspectives on NGO finance drawing on experiences with micro-enterprise credit. The 'radical' perspective sees the major finance constraint facing low-income households as related to the distribution of capital and access to other resources. This perspective pays particular attention to the need for NGOs to address poverty through finance and through changing social relations to enhance empowerment and justice. The 'alternative' perspective regards the constraint as a lack of both resources and consumption aspirations. Again, there is a concern with empowerment, but the alternative perspective places greater emphasis on community and collective rather than individual change. Both perspectives argue that programmes need to include some element of subsidy finance, in order to have a direct impact on poverty-reduction according to the radical perspective or as a moral issue according to the alternative perspective.

By contrast, the third perspective identified by Copestake, the 'minimalist' perspective, regards the main constraint to be the lack of access to financial services due to market imperfections. The rationale for this approach is the belief that poverty is not related to a shortage of financial resources *per se*, but to difficulties in gaining access to savings opportunities and loan finance. Indeed, some observers argue that the main constraint faced by low-income households is not the cost of financial services as many households are already paying above-market rates when they resort to moneylenders (M.S. Robinson 1996).[4] Under conditions where access is the constraint, the use of subsidies to reduce the cost of finance is seen as unnecessary, serving only to decapitalize the net value of the portfolio and to jeopardize financial sustainability (Rhyne and Otero 1992, von Pischke 1996). NGOs, therefore, are being urged to charge interest rates and fees which are sufficient to cover operational costs and to concentrate upon the delivery of finance-only services rather than offer a range of complimentary marketing and production supports (Dichter 1996, Rhyne and Otero 1992, 1994, M.S. Robinson 1996).

Advocates of the minimalist position point to the ability of micro-finance programmes to scale up and professionalize, and to cover extensive geographical areas in their operations. The Grameen Bank, for example, now lends over US$ 250 million a year and has a membership in Bangladesh of 2 million households. In Bolivia, BancoSol has over 70,000 members (70 per cent of whom are women) and the umbrella NGO ACCION claims 275,000 clients and has plans to increase this number to 2 million. One of the strongest claims for scaling up finance programmes is that the unit costs of administration fall as the number of loans increase (Rhyne and Otero 1994). As the president of ACCION, one of the largest micro-finance lenders, has claimed, in 'most humanitarian efforts . . . the cost of reaching every additional person brings the program closer to its economic limits. Successful micro-finance, on the contrary, becomes more self-sufficient with scale' (Microcredit Summit 1997a: 11). Thus, whereas in the past, a significant problem faced by programmes was the need to secure continued donor finance to pay for subsidized credit, the 'minimalist' position emphasizes non-susbidized loans to secure financial 'sustainability', generally interpreted as the ability of the agency to continue indefinitely into the future.

The strategies adopted to secure 'sustainability' have focused on reducing subsidies, particularly by accepting that market rates of interest will allow NGOs to attract additional (*private*) capital to the sector. Thus, over time, the argument that funds should be acquired through member deposits, in order to contribute toward mutual accountability and risk, is giving way to more conventional financial operations. There is a growing appreciation that in the conventional micro-finance model the funds revolve too slowly to deliver services on a scale sufficient to reduce operating costs substantially. Many programmes, notably in Latin America, have undergone a transition toward operating as formal banks and employing increasingly sophisticated financial instruments in order to secure renewable sources of funds (Hulme and Mosley 1996, Rhyne and Otero 1992, M.S. Robinson 1996).[5]

While the 'minimalist' position is rapidly becoming the accepted wisdom in some development circles (Copestake 1996) and it would appear to be increasingly difficult to obtain donor finance to work in other ways (Agency for Co-operation and Research in Development (ACORD), personal communication), a number of specific concerns about micro-finance in general and the minimalist position in particular have been raised. Some research, for example, indicates more moderate and less easily quantifiable impacts from microfinance than claimed by proponents (Berger 1989). While it is deemed successful at raising some households out of poverty, analysts suggest that micro-finance has helped the better-off poor and has not been an effective means to address structural poverty (Hulme and Mosley 1996, Johnson and Rogaly 1997, Montgomery 1996). In addition, there is scepticism that low-income households possess the capacity to sustain debt over the longer term, a feeling partly supported by reports of NGO staff using additional funds to assist households to maintain repayments during the early stages of programmes (Jain 1996, Montgomery 1996). Doubt has also been expressed as to the effectiveness and ethics of peer group pressure as a means to maintain repayment rates. Indeed, some NGOs, such as K-REP in Kenya, have found group lending to be of limited use and regard peer pressure to be an unfair delegation of lender responsibility to the borrower (Mutua *et al.* 1996: 188).

With or without this pressure, questions have been raised as to whether micro-finance programmes are capable of achieving high rates of repayment (typical claims are 90–98 per cent) unless also present are a high degree of supervision or hidden subsidies (Ebdon 1995, Yaron 1992). Others have gone a stage further and questioned whether NGOs possess the professional expertise to administer loan finance (Abugre 1993). Schmidt and Zeitinger (1996b), for example, found poor performance for all NGOs regardless of whether they operated according to a minimalist approach. Only one NGO from fifteen managed to operate according to financially sustainable criteria, despite an average real interest rate for the sample of 45 per cent which was partly a reflection of high staff-loan fixed costs.[6]

A specific area of debate has centred on the apparent ability of micro-finance programmes to scale up. One concern is that the rationale for operating 'minimalist' non-subsidized finance programmes is driven less by a belief that such programmes have the greatest impact upon poverty than by an acceptance that sustainability is now an essential attribute for NGO programmes (Dichter 1996, Rogaly 1996). Others have noted a contradiction between the ability of micro-finance programmes simultaneously to scale up and to maintain the quality of their portfolio. As programmes expand the expectation is that each new member is likely to present a higher debt risk than existing members, thereby reducing repayment rates and raising some administration costs. Observers argue that sustainable NGO lending may only be possible with substantial institution-building as a prerequisite so that increased scale does not jeopardize the close contact between NGO and community, the transparency of operations and the network of support achieved through a locally managed model (Schmidt and Zeitinger 1996a, von Pischke 1996).

Superficially, the debates on micro-finance outlined above are relevant to those NGOs that have either extended existing practices into housing finance or have been established solely for this purpose. Indeed, many housing and micro-enterprise finance programmes appear to be similar, especially where the former have built upon the traditions of informal-sector finance and NGO micro-enterprise programmes. As with micro-finance, many housing finance programmes use group-based collateral, the successful repayment of small loans as a precondition for larger scale finance and compulsory savings as a preparation for borrowing. As with many micro-finance programmes for enterprise development, NGOs are concerned to address the gap caused by the reluctance of formal-sector institutions to service the poor. However, a closer inspection of housing finance programmes identifies some fundamental differences in their organization and, more fundamentally, their analysis of the major constraints on development and therefore the strategies that they consider to be most appropriate to address poverty. Among NGOs involved in housing finance there remains a preference for the non-minimalist perspective.

MICRO-FINANCE, HOUSING LOANS AND
POVERTY REDUCTION

There are a number of obvious differences between the organization of finance for micro-enterprise development and that for housing, particularly if 'housing' in this context means construction. First, although small by comparison with average loans from formal institutions, housing loans are generally larger than those for enterprise development or income-generation and therefore imply higher levels of risk and/or longer amortization periods.[7] Second, especially for those NGOs that specialize in housing finance, the possibility of raising capital from private markets in the short term is limited as the sums required are relatively large.[8] Third, unlike successful micro-enterprise investments, housing improvements may not result directly in an income to the borrower that can be used for making repayments without additional investment. Instead, the greatest benefit may be in terms of savings in the cost of repairs and renovations as house consolidation takes place. With access to finance, households can afford to purchase a much higher quality of housing and to waste fewer resources. The cheapest housing materials are often scrap or non-permanent materials which require additional funding over time (and often annual purchases) for repair and replacement, despite the availability of better-quality alternatives. According to estimates made in India, over twenty years, pavement dwellers spend as much on repairing their shack as they would have to pay in loan repayments for a single room apartment (Patel this volume). Nevertheless, the less direct increase in disposable income means that financial management has to be both flexible and skilled if repayments are to be made successfully.

Of course, housing loans may raise incomes by permitting the construction of rooms for rent or a home-based enterprise. In Sri Lanka, for example, SEWA and the Women's Bank have found that loans for housing and neighbourhood improvement increased the number of small businesses such as small shops and tailors even in areas where they seemed previously to have barely existed. The women report that the upgraded houses help make a successful business by attracting more customers. The Women's Bank also found additional benefits: that food producers had increased space for food preparation, improved ventilation and enlarged counter space; garment producers could create space for private fittings, storage, production and design; shop owners had enlarged windows for trading and wider footpaths (Albee and Gamage 1996).

A more fundamental set of differences can also be discerned from an analysis of NGO housing finance programmes especially when these have not been established as extensions to existing micro-finance programmes. Although the efficient use of finance (and other resources) is considered to be important, most programmes subscribe to a view of development that includes recognition of structural problems and multiple disadvantages for those with low incomes, and that a primary consideration of such programmes is the need to redistribute income and/or assets in favour of low-income groups. Without some distribution mechanisms in place, the perception is that the beneficiaries will be those individuals and groups with access to investment resources, that is, the richer groups. In this context, housing savings and loan activity tend to be used as a means to strengthen local community organizations to represent better the interests of low-income groups, often through recourse to collective loans. In short, the principles underlying many NGO housing finance programmes are inconsistent with the 'minimalist' position that the problem of poverty can be resolved through more efficient markets.

Within an individual community, an effective programme can combine four identifiable advantages in a multi-faceted approach to poverty reduction. The first advantage is a material improvement in housing, infrastructure and tenure, and if combined with support for income-generation, to contribute to local economic development. A relationship between housing investment and the strength of the local economy has long been recognized. Through making housing improvements in conjunction with measures to increase the capacity of building materials producers, a financial injection into the community can be made more effective by ensuring that it circulates through the community two or three times. The housing improvement process can also offer an opportunity to teach local residents new skills and provide additional space for home-based enterprises, thereby strengthening livelihoods and bringing additional income into the local economy (Albee and Gamage, Patel this volume).

Second, housing finance programmes can support informal networks through the strengthening of local community ties and can make an important contribution to the reduction of the worst effects of poverty through sharing of both financial and non-financial resources, as well as the exchange of information and mutual solidarity (Trialog 1995). These networks are particularly important for women as they manage the multitude of tasks involved in earning income, caring for the home and dependants, as well as house construction. The strengthening of technical and financial skills, and the practice of organizing 'public' group meetings, can have important consequences for women's self-confidence and command of resources (Bolnick and Mitlin, Cosgrove, Datta this volume).

Third, housing finance can serve to strengthen the community skills needed to engage effectively with external agencies, regarded as essential if low-income settlements are to address their development needs. Certainly, housing finance management strengthens internal networks and collective action and this is a prerequisite for dealing effectively with external agencies (Bolnick and Mitlin this volume). Housing finance also offers community groups the practical experience of financial management and allows the development of financial trust, both of which might be necessary to deal with development projects and programmes. In some cases, housing finance management can provide a forum through which other collective priorities and strategies can be identified and developed (Bolnick and Mitlin, Patel this volume).

Finally, for the NGOs, housing finance can offer an efficient way to allocate development assistance (Arrossi *et al.* 1994). This can be achieved through a reduction in selection problems as recipient households are in part self-selected and lower levels of dependency as households contribute directly to the development process through the provision of 'counterpart' funding in the form of savings and loan repayments. The presence of counterpart funds also ensures that subsidies are allocated more efficiently because the households/communities become involved in the allocation decisions with some sense of ownership in the project. When collective finance management can also take place, transaction costs and risk management can be reduced, with the resulting savings to both the NGO and the borrower.[9]

In summary, then, the difference between minimalist micro-finance programmes and most housing finance programmes can be encapsulated by their attitude to subsidy finance and strategies for poverty reduction. Within an approach based on developing the efficiency of financial markets, subsidies are considered unnecessary and even detrimental to the effectiveness of such programmes: unnecessary because the major problem is a lack of credit not its cost; and detrimental because the subsidy is attractive to higher-income groups who try to obtain funds through the programme. Some multilateral agencies and NGOs seeking to improve housing finance provision support such a thesis (Diamond and Lea 1995, UNCHS 1991b). However, for those that adopt a social development approach subsidies are perceived as an integral component. Indeed, one of the rationales for housing finance is that it offers a better way of allocating financial support to low-income households and communities.

With a clear commitment to social development and working in a context of growing need, the more innovative NGOs have found themselves struggling to scale up the quantitative impact of their work. In a few countries with a favourable political context, housing finance programmes have sought to scale up their operations through working with government agencies. The following section discusses some of these experiences.

SCALING UP THROUGH GOVERNMENT PROGRAMMES

NGOs working on urban issues have long had to manage an interface with government agencies (Turner 1988, HIC 1997). In some cases, this interface has been difficult to establish and maintain, with the NGOs collaborating with government on urban upgrading and opposing government on evictions (Mitlin and Satterthwaite 1992, Institute of Housing Studies 1992). Moreover, for some NGOs, the idea of working with government has been difficult to reconcile with an ideology of independence, a concern to strengthen grassroots initiatives and in the face of perceived government hostility (Bratton 1989). The 1990s have witnessed increasing opportunities for effective collaboration either due to funding shortfalls, attitude shifts motivated by democratization, and concerns for decentralization and improved governance (Porio 1997). NGOs and donor agencies too recognize that working with government might permit them to lock in achievements and to improve project or programme implementation (Edwards and Hulme 1992b, Espinosa and López 1994, Mumtaz 1995, UNCHS 1993). Specifically with regard to housing finance, in a number of countries NGOs have sought to develop strategies for securing the wholesale adoption of their innovative strategies through government programmes (see Box 1).[10] All have drawn on NGO innovations and experiences, and employed former NGO staff; but they have also been planned on a scale that is much larger than any single NGO initiative.

Innovative government programmes

Philippines: the Community Mortgage Programme was established in 1988 to help low-income urban households acquire land titles and develop sites and housing. The programme provides loans to allow community associations to acquire land on behalf of their members, improve the site, develop individual titling of the land and provide housing loans for improvements or house construction. The associations select an 'originator' (an NGO, public or private agency) that serves to facilitate the technical work and guarantee the loans. As of September 1995, the Community Mortgage Programme had assisted 456 communities (55,218 households) with a mortgage value of US$ 49 million.

Thailand: the Urban Community Development Office was launched in March 1992 when the government approved a budget of US$ 50 million to initiate an Urban Poor Development Programme. The objective of the programme is to strengthen the capacity of the urban poor and those living in illegal settlements to obtain increased and secure incomes, appropriate housing with secure rights, an improved environment and better living conditions. The programme offers loans for income generation, small revolving funds and housing. By 1995, up to US$ 20 million of housing loans had been granted to 160 organizations and 15,000 households.

Brazil: *Mutirão* (mutual aid) was introduced in 1986 by the government in Brazil with the objective of providing 400,000 dwellings by 1990 when federal support was to end. Although this figure was never reached, in Fortaleza and São Paulo the programme continued with funds from the state and municipal authorities. Communal societies have been established throughout Fortaleza with representation from non-organized communities and federations of CBOs, unions and neighbourhood councils. Between 1987 and 1993 these societies constructed 11,000 houses in the city and an additional 3,000 in the remainder of the state. In São Paulo, the programme was enthusiastically implemented between 1989 and 1992, by which time 10,800 houses had been constructed and upgrading projects in more than 120 settlements had been started.

Sources: Cabannes (1997), Denaldi (1994), Prisliha (1995), HIC/ACHR (1994)

One reason is that these programmes are able to address the most significant problems faced by housing finance programmes. Perhaps the most critical constraint has been to secure the capital required to finance expansion. A number of different strategies have been used. In Colombia, the Fundación Carvajal has drawn in private-sector housing finance, in Chile NGOs have acted as guarantors of housing loans offered through a local bank, and in India and the Philippines, Northern NGO guarantees have enabled local institutions to obtain finance (HIC/ACHR 1994). All of these strategies, however, have proved difficult to put in place, whereas collaboration with government allows an immediate injection of funds.

Furthermore, these programmes have sought to replace the concept of 'financial sustainability' with one of 'political sustainability' (Somsook Boonyabancha, Urban Community Development Office, Thailand, personal communication). 'Financial sustainability' refers to a situation in which the future of an agency is secured because of its ability to cover all costs, including the cost of new capital, without recourse to sources other than commercial financial markets. 'Political sustainability' means securing the future of an agency through the distribution of subsidy finance (and other resources) in ways that strengthen the community organizations so that they are able to put pressure on the state or other institutions in order that the subsidy finance can continue. The strengthening of the community organizations' capacity to interact with external agencies is seen as critical to securing the continuation of the housing finance programmes.

These programmes have been significant in providing housing finance to many tens of thousands of households as well as strengthening local networks, and increasing the capacity of community organizations to interact with external agencies and attain reasonable repayment rates. While the scale has been limited relative to need, the programmes have reached many more households than most previous government or NGO programmes.

Despite these successes, a number of problems have emerged where governments have adopted innovative housing finance programmes. While this chapter does not offer a systematic analysis, we identify and explore three fairly consistent lessons to emerge from experience.[11] The first of these problems is that government bureaucracy has proved more resistant than expected to working in new participatory ways with the residents of low-income settlements. The experience of NGO-government housing finance programmes has to be placed within the constraints imposed by government. As a general review of NGO-government strategies concluded, collaboration usually involves 'poorly resourced, poorly motivated, usually bureaucratic agencies that are resistant to change' (Edwards and Hulme 1992b: 212). The second lesson is that the expansion of support services and, in particular, of technical advice and assistance has been insufficient to keep pace with the people's enthusiasm and the loan allocation process. The third lesson is that government priorities change and the pressure from social movements and people's organizations has not been sufficient to maintain these programmes in their original form. Below, we provide some comments on each of these three observations.

Working with government agencies

A major problem faced by NGO-government housing finance programmes is the interface between the state and the low-income community and, in particular, the level of bureaucracy associated with government institutions. Despite attempts to simplify procedures, experience suggests that problems with bureaucracy remain a significant deterrent to community involvement. In the case of one NGO-inspired government programme, the Fondo Nacional de Habitaciones Populares (FONHAPO) in Mexico, despite repeated simplification of bureaucratic procedures, community organizations have argued that these remained too difficult and were often compounded by the inclusion of additional agencies in the programme (Ortíz 1995). So complex and protracted did these procedures become, that some low-income communities have returned to illegal invasions and informal processes in preference to compliance with the formal ones (Enrique Ortíz, personal communication).

A similar picture emerges for the Community Mortgage Programme in the Philippines which obliged the NGOs to think about new fundraising strategies in order to cover the increasing costs due to delays in processing applications (Francisco Fernandez, Community Workers Programme, personal communication). The Philippine NGOs also found themselves having to raise additional funds (generally from Northern NGOs) because the US$ 20 per household to cover the costs of their support for the communities was insufficient and initial cost estimates had miscalculated the time that NGOs would need to commit to the programmes. The average time taken to process a community loan, from taking the first step to receiving the advance from government, is now two years compared to an original target of forty-six days (Community Mortgage Bulletin 1995a).

Of course, such delays are not unusual in more conventional programmes. Applicants to the Basic Housing Programme of the Chilean government have to wait over four years between registration and access to a house, and households wishing to undertake collective development face additional delays due to the problem of finding land (Rojas and Greene 1995). However, for programmes putting considerable emphasis on the collective involvement of local residents, such delays are of particular concern as, once residents are ready, projects need to be implemented rapidly.

Support services

In innovative housing programmes, loan finance is distributed directly to communities with technical assistance provided (and part or wholly financed) through a range of external agencies of which the largest group has been NGOs. To date, the experience of government programmes in low-income housing finance is that such programmes may have been too dependent on NGOs and on their capacity to support the process. There are often simply not enough sufficiently experienced NGOs to support the scale of community initiatives that emerge in response to such programmes (see Desai 1995 for Bombay).[12] In Manila, one research programme estimated that most NGOs concentrate on ten of the largest squatter settlements among the 2,000 such settlements throughout the city (Ann Dizon, personal communication). In Mexico, where NGOs have long been among the most innovative in working on housing issues, few of the 800 communities to obtain housing loans through FONHAPO received a high level of input from NGOs (HIC/ACHR 1994).

Yet there are few alternative institutions to NGOs. In one Mexican project, the community was divided into several groups: the one using an NGO successfully completed the programme, but the community being helped by a group of university students never finished their new homes. Even where the necessary skills do exist, NGOs may be reluctant to over-commit themselves to government programmes. The Centro de Vivienida y Estudios Urbanos (CENVI) one of the largest NGOs in Mexico and which has worked with FONHAPO on fifteen housing projects between 1980 and 1992, concluded that its experiences of 'informing, convincing, training and, eventually, accompanying the grassroots organisations themselves along the difficult route of low-cost housing development has taken up much of [our] time' (Connolly 1993: 74).

The absence of suitable technical support has implications for the linking of formal financial institutions with community financial institutions in order for such initiatives to be scaled up effectively (Mehta 1994). Recent programmes have tried to respond to this constraint. In Fortaleza, a number of complementary training programmes for different levels of professionals have been established in order to reduce the problems of inadequate technical assistance. In addition to courses at the university and technical school, training has also been provided for community leaders (Cabannes 1997). In South Africa, a programme drawing on many of the experiences described here is investing heavily in training programmes based primarily around community exchanges between experienced and less experienced settlements. Technical expertise is being provided both through professionals and local residents with construction skills (People's Dialogue and SPARC/NSDF/Mahila Milan 1994).

Maintaining programmes

Innovative government programmes for housing finance have been successfully established and have reached many thousands of households. However, experience suggests that such programmes have found it difficult to maintain their momentum and continue their development. For example, in 1986, the federal government in Brazil launched the *Mutirâo* (mutual aid) programme with the aim of building 400,000 dwellings by 1990 through collective loans. Nothing like this target was achieved before the federal support for the programme was ended and it only continued in Fortaleza and São Paulo because of the interest of provincial and municipal authorities. In general, there has been a reduction in the scale and content of these programmes. A number of specific difficulties have emerged to explain this trend. Some of the reasons reflect the general nature of the political process, others are related more to recent global trends. They include reductions in government expenditure that may translate into a reluctance to maintain funding for innovative projects, politicians wishing to influence and control successful programmes for political purposes, and pressure to 'privatize' collective finance.

Working within existing government institutions has proved difficult for several of the institutions considered here. The Community Mortgage Programme made a deliberate choice to locate itself within conventional government housing finance institutions to increase access by informal housing producers to formal sector funds. However, even from the outset, existing staff had little incentive to nurture and develop a social programme. In 1995, the government committed itself to allocating just under US$ 500 million to the Community Mortgage Programme between 1995 and 2002 but timed the allocation so that about 70 per cent of this finance will be available only in the last four years. After allocating US$ 26 million in 1992, commitments have fallen to an estimated US$ 9 million in 1995, raising concerns that the full funds will not be forthcoming (Community Mortgage Bulletin 1995b). Despite a campaign by NGOs involved in the programme, it has proved difficult to secure a stable government commitment and additional finance.

A change in the political administration resulted in the interruption of a housing finance programme in São Paulo called Fundo de Habitacao Sub-Normal – Comunitario (FUNACOM) (Denaldi 1994). This programme was the response of the municipality to co-ordinated pressure from the Union of Housing Movements in São Paulo which had been active throughout the 1980s in a number of mutual aid housing projects that served as a pilot for FUNACOM. When the new administration interrupted the programme, the union organized campaigns and demonstrations, but without much success. The union is now attempting a dual strategy of continuing with land invasions to put pressure on the administration and negotiations with the state government to seek new sources of support for self-managed housing development (ibid.).

Finance programmes can become victims of their own success as they become vulnerable to politicization. In Sri Lanka, the repayment rate for the Million Houses Programme fell from 65 to 6 per cent after a government decision to write off payments for members holding food stamps (Anzorena 1993). Alternatively, programmes have been restricted or attempts made to co-opt the NGOs for representing a danger to established political interests (Vakil 1996). This threatens to make the NGO a less vociferous and effective advocate within government and may have consequences for the reputation of the NGO with low-income households and with other organizations.

In recent years, there has been a general tendency for governments to withdraw from many public services including direct involvement in housing and housing finance. To a degree, the innovative government and NGO initiatives described here have managed to combat this trend. In part, this reflects the increased ability of NGOs and people's organizations to demonstrate the strength of their case and negotiate with government to obtain resources. However, current ideological trends are also evident in the programmes. First, there has been a tendency to individualize housing finance with loans being held by the household and not the community. In the Philippines, there is now pressure from the government to individualize community mortgages into individual titles within two years of credit initiation. A similar situation is being faced by the traditionally strong housing co-operative movement in Uruguay which is subject to government policies that undermine collective organization (Miranda and Frigerio 1994). This trend means that few of the benefits to community organization for development can be realized. Second, there has been a greater emphasis on private-sector production rather than reliance upon local community enterprises. Inevitably, this increases production costs. Third, there has been continued pressure to reduce the value of subsidies. In order to achieve this, governments are trying to target subsidies more accurately towards the poorest groups. As a consequence, there is a reduction in the scope of programmes in Mexico, Brazil and the Philippines, although it is not evident that the poorest groups have been protected.

The experiences of the past twenty years have included a number of innovative NGO housing finance programmes developed in response to the needs of low-income communities that have successfully assisted residents to improve their housing. In response to these innovations, and under political pressure to address the needs of low-income households and demands of urban social movements, governments in a small number of countries have been encouraged to establish programmes that have drawn heavily upon NGO experiences. These programmes have reached large numbers of households, albeit with a number of unanticipated problems relating in part to the ability of government institutions to work with low-income households.

CONCLUSION

This chapter has described some of the innovations in housing finance for low-income households that have taken place during the last twenty years, with particular reference to the work of NGOs. While we have been critical of certain aspects of NGO performance, there is no doubt that such programmes can play an important role in addressing housing needs and poverty alleviation. To paraphrase Schmidt and Zeitinger (1996a: 248), the question is not whether NGOs should be instruments of development, but how they are to be incorporated.

At present, NGOs involved in housing finance come from two positions. On the one hand, NGOs primarily established to provide loans for micro-enterprises are showing an interest in extending their lending to housing. On the other, a number of NGOs which have long been working in the field of urban development have sought to provide access to finance, in some cases using savings and loans. We have argued that the design of these housing finance programmes has to be seen in relation to emerging development finance philosophies (broadly: radical, alternative and minimalist) which translate into different approaches with regard to the role of subsidies and the perceived relationship between financial services and poverty reduction. For the most part, housing finance programmes have avoided the minimalist approach, regarding housing as part of a broad agenda that includes poverty reduction, community building and resource distribution. While this approach has implications for long-term financial sustainability, NGOs have argued that their effectiveness should be gauged according to their impact upon the livelihoods of local residents and communities, the strength of the local economy and the ability of communities to engage with government as well as housing construction.

Consequently, NGOs have adopted a non-minimalist strategy towards scaling up housing finance programmes. NGOs have argued that in order to address the multiple needs of low-income groups for loan finance requires some level of subsidy, innovation with micro-finance for income generation and housing, and substantial additional funds to scale up programmes. Experience shows that international development assistance is unable to provide sufficient capital to meet the scale of need. The alternative of seeking commercial capital and developing micro-finance lending can help to reduce poverty and address broader development needs, but is likely to have two outcomes. First, the lowest-income groups are unlikely to be reached if only commercial funds are available. Second, short-term loans may curtail the possible investment in improved housing even if they are made incrementally.

This leaves engagement with government, either in collaboration or as part of the government structure, as the most viable approach. Although by no means guaranteed, programmes can permit NGOs to draw upon government resources and technical skills to address problems of quantity, without unduly compromising the NGOs' comparative advantage in providing qualitative resources. To date, evidence suggests that to realize this strategy successfully is a difficult balancing act. If NGOs are to secure government support for housing finance programmes offering collective loans to low-income communities, then a number of factors need to be met. In particular, the programmes need to gain the support of the communities (or social movements) themselves in order for the pressure to be maintained on the government to continue programmes. They also need to judge carefully how they engage with government if they are to avoid the risk of bureaucratization and politicization. In many cases, NGOs lack the capacity and resources to conduct these tasks.

The experiences discussed in this chapter suggest that by working with NGOs and community-based organizations, governments can better address the housing needs of low-income groups. What is required from both NGOs and governments is an honest acknowledgement of existing expertise and capacity. To assist this process requires an appraisal of existing programmes to be carried out. The chapters in this volume can contribute to an informed discussion of how future housing finance programmes can meet the housing needs of low-income groups. But, we also recognize that much more analysis and experimentation is needed, especially at the local level, before governments, NGOs and international development agencies can understand how to support the process of housing construction and improvement that, at the current time, is mostly financed by the low-income households themselves.

NOTES

1 According to Satterthwaite (1995) urban expenditure accounted for less than 3 per cent of development assistance between 1980 and 1993.

2 For example, in Rio de Janeiro, a Brazilian NGO working in housing upgrading and rights has extended its work into housing co-operatives and loan finance and, in the Philippines, an NGO working on evictions and land tenure issues has become involved in a loan programme for infrastructure once land leases have been obtained.

3 The use of phased or graduated loans is particularly important to housing finance due to the relatively larger sums involved. Part of the rationale is also to promote savings with each subsequent loan containing a higher proportion of self-finance and a lower proportion of subsidy.

4 In India, when communities were allowed to decide the rate of interest on loans they often chose levels which NGOs thought to be unaffordable (Mehta 1994).

5 BancoSol is part-financed through the placement of certificates of deposit on US stock markets, the Fundación Paraguay de Cooperación y Desarrollo has issued US$ 150,000 on capital markets and ACCION is set to increase its lending from US$ 1 billion 1992–7 to 11 billion by the year 2005 by supporting local NGOs to issue bonds on local markets (Microcredit Summit 1997a: 30-1, 1997b: 17).

6 It is appreciated that micro-finance programmes and NGOs incur particular costs, but the figures produced by Schmidt and Zeitinger compare badly with the management charges reported by Boleat (1987: 161) for building societies.

7 Most programmes have loan periods of between three and seven years, periods that are significantly longer than those for micro-enterprise loans.

8 Even in Latin America and Asia where NGOs have a longer track record with private

capital markets, there are few examples of using this route as a source of funds when the purpose is to extend housing finance.

9 Bennett *et al.* (1996) quote an International Fund for Agricultural Development (IFAD) appraisal of a rural NGO project in which local group management involved savings of 30 per cent for the lender and 75 per cent for the borrower in comparison with a situation where a bank lent directly to individuals.

10 For reasons of space we elaborate here upon just three agencies. Others include Fondo Nacional de Habitaciones Popularest (FONHAPO) in Mexico, Fundación Promotora de Vivienda (FUPROVI) in Costa Rica and Programa de Desarollo Local (PRODEL) in Nicaragua (Ortíz 1995, UNCHS 1993).

11 This section draws upon Mitlin (1997).

12 A survey of NGOs in urban Africa found that about two-thirds were involved in education, almost one-half in health provision and welfare, while less than one-fifth had any involvement in housing (Mazingira n.d.).

3

HOUSING FINANCE IN DEVELOPING COUNTRIES

A review of the World Bank's experience[1]

Robert M. Buckley
The World Bank, USA

INTRODUCTION

Since 1983 the World Bank has lent almost US$5 billion (1996 dollars) to support housing finance in forty projects in developing and transition economies. Of the twenty-seven projects that have been subjected to *ex post facto* evaluation twenty, or 74 per cent, were found to have had a satisfactory outcome, a figure somewhat higher than the 68 per cent realized by other World Bank projects, and considerably higher than the performance of loans the World Bank has made to financial intermediaries.

This chapter attempts to describe these projects, the rationales for undertaking them, and some of the factors that caused the relatively stronger performance of housing finance interventions *vis-à-vis* those of the Bank in other sectors. It provides a simple analytical perspective that links these projects to the World Bank's evolving understanding on how to assist the development of effective and sustainable financial and housing delivery systems. Particular emphasis is given to the effects that the changing world financial and economic environment has had on the kinds of projects undertaken. Finally, the conclusion argues why housing finance reforms should be an important part of the development process. It might be appropriate to mention at the outset that the emphasis in this chapter is on housing finance projects in a country development strategy. Little attention is given to the important micro-economic aspects of such lending. Nor does the chapter deal with community or NGO finance schemes, important though they are in providing housing for shelter.

HOUSING FINANCE AND FINANCIAL
SECTOR PERSPECTIVES

The housing finance delivery mechanisms observed in developing countries are often complex institutions that are structured in a surprisingly wide variety of ways. This variability in institutional structure occurs for a number of reasons. One important reason is that in many cases these institutional structures are, in effect, the results of the accumulation of the accidents of history. Issues such as the rate of urban–rural migration, the level of per capita income when this migration takes place, the nature of the financial system when urbanization trends intensify, and even the exchange rate policy pursued, all affect the kinds of housing investments undertaken and the way they are financed.

It follows that the evaluation of these institutions is very country-specific and idiosyncratic, akin perhaps to a form of financial anthropology. Such a field of study, unfortunately, does not lend itself to easy generalizations. Thus, at first glance, it would appear that analysis of housing finance institutions would be largely a taxonomic exercise without much strategic perspective possible. Fortunately, despite the significant differences in structure across countries, there are also strong underlying similarities across groups of countries. As a consequence, some common evaluative principles can be developed which can be applied in most circumstances.

To give a sense of these principles this chapter presents guidelines on housing finance presented by a recent World Bank policy work, *Housing: Enabling Markets to Work* (World Bank 1993a). These guidelines provide a summary of the components of an effective housing finance system. Indeed, these components can even be enumerated into a table of 'dos' and 'don'ts' of effective policy. Obviously, this list of principles is meant more as a leitmotif of the elements of a sustainable system than as a prescription or set of rules. Nevertheless, it does contain the basic building-blocks of a sustainable system regardless of that system's institutional structure. To make this argument more concrete it is necessary to review the evolution of the World Bank's involvement in housing finance and the lessons learned from this experience.

EVOLUTION OF THE WORLD BANK'S HOUSING
FINANCE PORTFOLIO

World Bank lending for housing now operates in an environment that has changed in basic ways over the almost twenty-five years since such operations began. No longer is the Bank lending only to the lowest-income countries undergoing rapid urbanization. With the debt crisis in 1982 came a recognition of the importance of improving domestic resource mobilization capabilities. As a result, the type of projects financed became very different. The types of borrowers changed from almost exclusively public-sector institutions to financial intermediaries and the loans changed from being small demonstration projects to larger loans. Instead of an almost exclusive focus on the physical characteristics of the assets financed, attention shifted to the institutional structure of the implementing agencies and their ability to mobilize and manage resources. In other words, attention shifted to the other side of the implementing agencies' balance sheet: more attention was given to the institution's liability structure than to the type of asset financed. Inducement rather than directive became the vehicle of support. In the urban sector a clear shift in the type of lending has taken place and this has been perhaps most pronounced in lending for housing.

This new approach gives considerable attention to the importance that policy distortions and local ownership can place on moving to more market-oriented systems. It is perhaps no coincidence that this change in approach intensified following the fall of communism, and the rise of democratically elected and decentralized governments throughout the world. To sum up, the urban public institutions involved in World Bank projects now are more likely to be accountable institutions developed by democratically elected constituencies. Similarly, the NGOs involved in implementing Bank loans are more likely to be private-sector firms, as was the case in shelter projects in Ghana and India. Both the public and private institutions involved are much more demand-responsive than their counterparts during the early years of Bank urban lending. These new Bank partners are also considerably more likely to be operating in a financial system that is more competitive and open to inter-sectoral resource flows. Finally, these institutions now operate in an environment that will compel them to be much more self-reliant.

In the years since the Bank's first housing finance project, more than forty projects focusing on housing finance or including major housing finance components have been approved, totalling almost US$5 billion (1996 dollars). Indeed, within six years of the first housing finance loan the volume of Bank lending for housing finance had already exceeded the total sixteen-year volume of sites-and-services lending since the first such project in 1972.

Over 80 per cent of housing finance projects were channelled to financial intermediaries rather than to non-financial public-sector housing authorities. In addition, in the cases where projects relied on government housing programmes, such as in Korea, Thailand, Chile, and Mexico, a central objective was a refocusing of the public-sector housing authorities so that the financial sector could more actively participate in the mortgage market.

While this refocusing of the public sector has taken a number of different forms, the central objectives have been to reduce the overall level of subsidies and to better target them on lower-income families. In Mexico, Chile, Morocco and Zimbabwe, follow-up housing finance projects carried this disengagement of the public sector from lending one step further. Most of the institutions that received housing finance loans have either been created in recent years or considerably refocused. Cumulatively, these borrowing institutions for shelter projects are now more likely to be financial intermediaries than public housing entities and they are likely to be newly created or recently restructured institutions.

THE EFFECTIVENESS OF HOUSING FINANCE OPERATIONS

Any assessment of the impact of housing finance interventions on the overall development strategy of the Bank will depend on the perspective taken. This section focuses on how well the lending operations of the Bank have addressed the following questions. First, is there a clear strategy for the reform and development of the housing finance sector that is consistent with broader financial sector objectives? Second, is the strategy consistent with the objective of restoring fiscal balance and controlling and targeting subsidies along the principles? Finally, does the strategy promote a well-functioning housing market and help low-income families get access to adequate housing?

In evaluating the financial sector implications of projects, emphasis is placed on project performance with respect to domestic resource mobilization efforts. The statistics are therefore similar to those presented in evaluations of financial intermediaries. Attention is given to the projects' effects on borrowers' interest rate policies, on their efforts to mobilize domestic financial resources and the response to loan arrears. In addition, the effect on the government's contingent liabilities is also examined. For fiscal policy concerns, attention is given to the level, transparency and targeting of subsidies and, again, the government's contingent liabilities. Finally, for housing sector concerns, the focus is on institution-building, the regulatory environment, and the unit costs of the houses financed by the programme. These concerns are less quantifiable than are the financial or fiscal policy measures, but no less important.

Financial sector objectives

For financial sector objectives to be met, the kinds of transactions that occur require that borrowers and savers freely enter into the contracts on a voluntary basis. To do otherwise is to penalize other borrowers or savers who will ultimately be discouraged from providing resources for further borrowing to take place. This view does not mean that subsidies are inappropriate. Subsidies and their effective targeting are the subject of the next section on fiscal concerns. Here the concern is with establishing approaches that can sustainably provide financial resources. Such an approach generally requires that savers be provided a return that exceeds the inflation rate on their savings, as well as a return that reflects the risks involved in getting their money back. Hence, positive interests rates (lending rates in excess of the inflation rate) are viewed as a prerequisite for successful, sustainable finance. In addition, savers must be able

47

to expect this return after the costs of paying off arrears is subtracted from the returns paid.

Resource mobilization

Almost 80 per cent of the projects attempt to promote savings mobilization and more than 60 per cent of the housing finance institutions that have received Bank loans have been able to mobilize most of their resources domestically. This result is in sharp contrast to the less than 10 per cent of Bank-supported development finance institutions that did so through the 1980s. The latter is also in contrast to the results of a sample of agricultural credit loans which indicated that only 12 per cent of the agricultural projects attempted to promote savings mobilization. The greater emphasis on institutional strength is certainly consistent with the findings, noted earlier, that the housing finance projects have generally out-performed those of other Bank loans through financial intermediaries.

The inflation-adjusted interest rates paid for funds mobilized at the time of the project appraisal were overwhelmingly positive, 90 per cent of the projects. These results are important for institutions which will ultimately be sustainable only if they can induce savers to place their savings with them. Whether these funds are mobilized directly or by other intermediaries and then lent on to the housing finance lender depends upon the type of financial infrastructure that exists. But, regardless of the type of system, positive interest rates are essential to provide long-term financial integrity to the intermediation process.

Risk exposure and contingent liabilities

Another increasingly important perspective on how projects affected the government's role in mobilizing financial resources is whether the project changed the government's domestic risk exposure. Governments are exposed to risks or contingent liabilities when they explicitly or implicitly support institutions that provide finance. One of the most famous recent instances of this kind of risk exposure are the costs realized by the US government in protecting the depositors in the savings and loan associations. Such risks are of particular concern in longer-term lending, such as housing finance, because the risk to the government is, of course, affected if it is a depositor in the housing finance institution or if it guarantees other depositors. Moreover, in practice, it also assumes contingent liabilities, or 'moral obligations' to make up losses, if it sets lending interest rates or regulations which undermine the viability of the housing finance institution.

Only a few of the projects increased the government's risk exposure with respect to domestic (as opposed to foreign exchange) contingent liabilities. In addition, of the projects that did increase the government's domestic contingent liabilities, other desirable objectives were usually achieved. This kind of loan, in effect, replaced a financing mechanism that ultimately conferred large credit subsidies with a mechanism that provides much smaller liabilities whose cost may, but probably will not, be realized.

Arrears

A number of features of repayment behaviour appear to be noteworthy. First, the average per centage of portfolio arrears reported for housing finance institutions is considerably greater than the rate realized by development finance institutions and, second, most of the projects appeared to have portfolios of commercial quality (that is, arrears of less than 6 per cent of the portfolio). In addition, according to simple measures of expected inflation, the real interest rate charged was in the order of 3.6 per cent. Finally, the apparent commercial level borrowing rate on so many of the projects standards in sharp contrast to the results for Bank-financed shelter projects in the 1970s. The average on-lending rate for shelter projects during the early periods was minus 3 per cent.

As the data indicate, housing finance loans can be made at positive real rates through reliance on domestic resources and with reasonable expectation for satisfactory cost recovery. However, it should be recognized that the development of more competitive and sound financial practices are long-term objectives. They may not be achievable in one project, particularly when housing sector and financial sector policies are at odds with each other or have been conducted through fragmented policies for a number of years. In many countries, lasting improvements in the functioning of these sectors will require an incremental approach as has been pursued in Morocco. Similarly, follow-up housing finance projects have been prepared in Zimbabwe, Chile and Mexico, and more are planned for reforming socialist economies such as Russia, China and Albania.

While the development of an ongoing relationship is important to success, so too is innovation. An example of a housing finance project that helped induce innovation is the loan to Ghana. This project provides for a new mortgage instrument that permits household debt to replace large implicit credit subsidies in a heavily indebted country. It provides for the index-linking of mortgage repayments to wages and recognizes that in an economy such as Ghana's, although real wage decreases may well imply temporary repayment problems for borrowers, they do not necessarily imply a reduced long-term ability to repay. Under the project, if real wages behave in such a way that repayments are not sufficient to amortize the loan, the shortfall can be capitalized into the outstanding loan balance. Simulations of various future real wage scenarios indicate that as long as inflation does not get out of hand, even with very pessimistic assumptions, the indexation scheme can eliminate subsidies.

If this instrument becomes more broadly operative than it already is, it is likely to be a good example of a financial innovation that helps both the urban sector and the financial system. It shows how the development of mortgage instruments that are in tune with macro-economic conditions can help ensure that lending institutions have competitive access to resources. These resources can be mobilized in reaction to demand and can help keep the need for subsidies to a minimum.

In sum, an obvious implication of moving to market-rate lending, as the housing finance projects appear to be doing, is concern for the target audience of beneficiaries. That is, as borrowing costs increase, low-income households may no longer be able to afford to participate in housing programmes. This result, in turn, raises the issue of how subsidy targeting or fiscal policy concerns are addressed in housing finance projects.

Fiscal policy concerns

Fiscal policy concerns were the primary objective of a quarter of the housing finance projects. These projects were concerned largely with the broader effects of housing subsidies on government expenditures and transfers. The Malawi project, for example, was initiated by earlier structural adjustment discussions and the Chilean project and the Mexican loans were components of broader fiscal policy dialogues between the Bank and the borrower.

While both of these Latin American country projects were directed at the fiscal aspects of subsidy targeting, an equally important dimension was their reduction in the implicit taxes on the institutions which financed these subsidies. In Mexico, commercial banks were required to allocate an amount on the order of US$ 400–700 million per year to fixed-rate low interest rate mortgages (see Siembieda and López this volume). The Bank loan financed a time slice of commercial bank mortgage lending that eliminated much of this implicit tax on the banks. The elimination of this implicit tax was the product of a dialogue between the Bank and the government of Mexico. High quality studies of this issue were produced by the Mexicans and provided a key input into policy discussion.

Subsidizing target beneficiaries

A number of the fiscal features of the projects were prominent: the target beneficiaries of housing finance projects are 'below-median income' households, rather than those with low incomes. More than one-half of the projects provide cross-subsidies to lower-income borrowers. One assumption seems to be that if lower- and middle-income households can be accommodated by debt at commercial rates, subsidies that had been going to them could then be redirected to those who truly need them. The lesson learned from the Bank's sites-and-service projects is that the production of low-cost affordable housing is possible. If the right housing standards are in place, it is access to finance, rather than concessional lending terms, that can best improve housing conditions.

Further, in the early years of lending for housing finance a number of supported institutions provided access to concessional finance for very expensive housing. In some countries, projects financed mortgages of up to US$ 17,000. However, the institutions which receive Bank loans could finance mortgages of up to US$ 50,000 at preferential interest rates. There is no rationale for these institutions to provide credit at preferential interest rates, which is exactly the approach that the Bank's *Housing: Sector Policy Paper* warned against (World Bank 1975).

Besides the subsidy targeting concerns associated with housing finance projects, an equally important concern is the efficiency with which housing transfers are mobilized. For example, forcing financial institutions to invest in below-market-interest rate securities which ultimately finance housing loans, as has been done in many countries, is to rely upon a very inefficient transfer mechanism. Such devices can, in the long run, seriously erode the ability of financial systems to mobilize resources. Greater subsidy transparency not only yields a better measure of who is the subsidy beneficiary, it also yields a better sense of how government policy is affecting financial development. In short, transparency requires that the subsidy element be calculated on a transaction and that it be treated as an obligation of the government. This kind of measure helps give a sense of whether the borrower's subsidy is borne by the government, other borrowers (through higher interest rates) or deposits (through negative interest rates).

Foreign exchange risk

Housing investments produce either no pecuniary income or rental income denominated in local currency. In addition, mortgage payments can make up as much as 25–30 per cent of household income that is earned in local currency. Consequently, in the case of housing it is not possible for the ultimate borrower to carry the foreign exchange risk. Over the longer term such projects should clearly encourage domestic resource mobilization to finance housing. In this regard, the projects in Zimbabwe and Lesotho which aimed for 100 per cent domestically-based resource mobilization had noteworthy aspirations: all housing finance was to be domestically mobilized.

In the short run, because housing finance institutions finance domestic assets they are not able to bear foreign exchange risk. Thus, normally, the government is willing to assume this risk for a fee. However, if the housing finance institutions, and, by implication, the final borrower, do not pay a fee for the transfer of this risk or pay a fee that is lower than the expected cost of this risk, the subsidy element should be recognized.

HOUSING SECTOR CONCERNS

As noted in Table 3.1, the most frequent rationale for housing finance projects is sectoral policy concerns. These concerns arise because of the inability of most housing production delivery mechanisms to accommodate the large and growing demand for housing. In almost every instance in which financial or fiscal policy concerns were the basic policy objective, these broader concerns were generated by side effects of the functioning of the housing and housing finance delivery mechanisms. Only in the cases of Malawi and Albania were the projects developed without prior housing sector work or extensive technical assistance studies.

Those projects in which housing sector objectives were identified as the principal rationale for the project can be subdivided into two sub-objectives. First, those that sought to demonstrate that many lower and moderate income households can afford to repay market rate finance. These efforts, in many respects, involve giving greater attention to the design and implementation of both effective mortgage and building practices. Second those projects which gave greater emphasis to the private sector in providing both housing and housing finance services. These efforts often seek to disengage the public sector from functions that the private sector can perform.

However, evaluating the sectoral performance of housing finance projects against these kinds of objectives is more difficult than is the evaluation of their financial or fiscal performance. It is more difficult because, as noted at the outset, the constraints on the development of more effective housing finance systems vary so widely depending on the characteristics of the economy. In addition, these kinds of constraints are difficult to quantify in simple summary statistics.

Nevertheless, even though these kinds of characteristics may be difficult to measure they are important to the development of successful housing finance projects. In an attempt to get a better sense of how policy constraints are affecting the sector the World Bank, together with UNCHS-Habitat,

Table 3.1 The chief rationales for housing finance projects, 1983–95

Financial sector reforms	Broader fiscal policy initiatives	Housing sector concerns	
Côte d'Ivoire	Chile	Albania	Nigeria
Ecuador	Chile II	Argentina	Papua New Guinea
India	Ghana	Fiji	Philippines
Mexico II	Malawi	Indonesia	Russia
Morocco	Mexico	Korea	Thailand
Senegal	Mexico I	Lesotho	Vanuatu
Uruguay	Poland	Mexico	Zimbabwe
	Russia	Morocco	

established the Housing Indicators Program to track a series of housing market performance indicators in more than fifty countries (World Bank 1993b). These indicators show the relationship between supply conditions, policy and the regulatory constraints on property use and development.

GUIDELINES FOR EFFECTIVE HOUSING FINANCE

The Bank is still in a transition from a traditional housing sector analysis, which looked at the demand side in terms of 'housing needs' and which emphasized the role of government as 'the' provider of shelter services, especially for low-income households, to a broader analysis that puts housing and government housing policies in a broader economic perspective. This shift in perception has been followed by a shift in lending approach. However, the transition away from sites-and-services and upgrading projects to include housing finance operations will not occur rapidly; nor, in many countries, should it occur at all in the short term.

Most of the Bank's early shelter loans were made to lower-income African countries that were urbanizing rapidly, had very basic financial systems and often had severe land rights problems. The average per capita income level of countries receiving a sites-and-services project was 40 per cent less than that of the countries receiving housing finance loans. In most of the lowest-income countries, the sites-and-services approach was the appropriate strategy then, and in many of these countries it remains the appropriate strategy today.

Almost every housing finance project was preceded by other kinds of shelter projects to build up the basic institutional infrastructure and in most of the borrowing countries new or refocused institutions are the borrowing agency. In many respects, then, housing finance projects represent a 'second generation' approach that should not be of the foremost priority for many of the Bank's lowest-income borrowers.

Ultimately, for the Bank, it appears that some regional specialization of housing finance strategies will almost inevitably occur. The type of housing project that is appropriate in one type of country will be inappropriate or of much lower priority in another. For example, in many higher-inflation countries better mortgage index-linking may be essential to make housing affordable, to reduce transfers and to help mobilize financial resources. In such an economy, housing finance interventions should be of relatively high priority. In a high inflation country that also has experienced a low level of economic and financial development, declining real income and a weak land cadastral system, there are almost certainly many more important sectoral policy issues to be addressed before indexation is introduced or discussed. Stopping the leakage of resources through negative *ex ante* interest rates is essential in all high inflation countries. But whether mortgage indexation is the appropriate means of doing so is another question.

With this historical detour in hand, consider the following guidelines for effective housing finance. Table 3.2 provides a summary statement of the 'dos' and 'don'ts' of financial policy as applied to housing finance.

Table 3.2 Dos and Don'ts in developing mortgage finance

DO	DON'T
Allow private sector to lend	Allow interest rate subsidies
Lend at positive interest rates	Discriminate against rental housing
Enforce foreclosure laws	Neglect resource mobilization
Ensure prudential regulation	Allow high default rates
Introduce better loan instruments	

Source: World Bank (1993a).

Of course, each phrase in the table covers a number of issues that may be very difficult to achieve in a particular country context. It is easier said than done, for example, to suggest that lending should be done at positive real interest rates when the entire financial system does not follow such practices. Similarly, how can foreclosure laws be enforced when there are none or it takes the court system ten years to provide adjudication? Nor is any guidance given about which practices should be given priority, for example, prudential regulation or private sector lending. Nevertheless, the basic building blocks of an effective system are *analytically* relatively straightforward even if the process of getting to such a system is one that may require considerable strategy.

The table illustrates that housing finance should be competitively supplied at market interest rates by prudentially-sound institutions. The institutions should operate in well-designed and transparent regulatory systems that provide adequate legal recourse for both borrowers and lenders. Inter-institutional lending should be based only on clear accounting standards and practices. Projects financed by these systems should be demand-determined and, to the greatest extent possible, seek full cost recovery. In cases where projects are unlikely to or do not achieve full financial cost recovery, a clear source of either subsidy funding or contingent risk-bearing should be identified at the outset. Finally, subsidies and finance should be identified and separated.

CONCLUSION

The process of moving towards market-based, competitive housing finance systems is in full train across developing and developed countries. New methods of finance are emerging, and as the processes of greater financial integration come into play, the methods of financing shelter will undoubtedly continue to undergo substantial change. This process of change will take many forms and will rarely come quickly. Nevertheless, it is worth remembering that the fits and starts of this process, as reflected in the stresses on existing institutions, are often responses to underlying market forces. It is also worth noting that establishment of a market-based system ultimately holds out more than the prospect of being able to use resources much more effectively. In some respects, the change in housing finance systems around the world is much like the replacement of an old car with a newer model. The old car simply will not work in the new environment. It is not a matter of if it will crash, but rather a matter of when. For safe driving in the new, high-speed environment, a car with a better design, not just a new car, is needed. Consequently, housing finance reforms should be pursued not only because they can provide efficiency gains and trickle down distributional benefits, but because they can help avoid serious economy-wide disruptions.

NOTES

1 The views expressed in this chapter are not those of the World Bank. This chapter is an expanded and amended version of chapters 11 and 12 of *Housing Finance in Developing Countries* (1996) Macmillan.

Part II

FORMAL AND INFORMAL HOUSING FINANCE

4

THE POLITICAL ECONOMY OF FORMAL SECTOR HOUSING FINANCE IN JAMAICA

Thomas Klak and Marlene Smith
Miami University, USA, and the National Housing Trust, Jamaica

INTRODUCTION

The 'dream home' of most Jamaicans is a house built in block and steel with at least two bedrooms and a garden. But, owing to a wide and increasing gap between income levels and house prices this aspiration has proven to be elusive as shelter costs have greatly exceeded wage increases. Consequently, although 70 per cent of Jamaicans own their homes, this includes shelter of varying quality, occupant density and security of tenure (Eyre 1997). In order to resolve the housing problem, the national debate has centred around the role of the state. From the 1960s to the 1980s, there was consensus among both economic and political elites that state intervention in the housing market was essential if shortages and inadequacies were to be addressed. Today, this consensus no longer holds and formal sector investment in housing has declined steadily as a share of GDP (which itself has grown only modestly). In part, this can be attributed to the unleashing of market forces through structural adjustment programmes with their emphasis on exports, foreign exchange earnings and privatization. Such free market sentiments have impacted upon public housing policies because of their consumption of scarce foreign exchange and have left little room for an interventionist state. Consequently, state activity in housing has been transformed in the neo-liberal era which assigns it the role of 'enabler' or 'facilitator'.

An organization which has survived this transition to privatization is the National Housing Trust (NHT), the state's main housing agency, which was established in 1976 with the clear mandate of promoting the housing needs of low-income households. The survival of the NHT in the face of both domestic and international pressures is due largely to the fact that it serves two critical constituencies: middle-income groups, especially those employed by the state, and the two major political parties, the Jamaican Labour Party (JLP)

59

and the People's National Party (PNP). In power, both of these parties have manipulated the NHT to pursue their own ideological and clientalistic agendas.

At present, the Trust is the largest and most powerful player in Jamaica's housing system. Endowed by a steady inflow of funds from compulsory payroll deductions, the NHT continues to be, in terms of its asset base, one of Jamaica's largest financial institutions. Indeed, by the late 1980s the NHT's mortgage portfolio was worth considerably more than that of any collection of private financial institutions such as credit unions or building societies. At this time, the NHT held 28 per cent of the country's mortgages by value and a much larger share by number considering that, on average, private institutions provide larger mortgages. To further illustrate its predominance, the NHT collects about 80 per cent of outstanding government mortgages by value (Peterson and Klak 1990).

Increasingly, however, the NHT has marginalized its (supposed) main constituents, low-income groups. An examination of the allocation of the NHT's financial resources reveals that over the last two decades, a person in the bottom 40 per cent of the formal work force has had just a 2 per cent chance of obtaining a mortgage. Low-income groups have not been powerful enough to sway the NHT significantly in the direction of the large-scale provision of low-income housing.

This chapter explores the organization and performance of the NHT in distributing finance in the context of the struggle for basic needs such as shelter, state socio-economic interventions, the role of regimes in crafting their own policy responses and neo-liberal transformation. By examining the NHT's funding base, expenditures and beneficiaries, the chapter outlines the scale of the financial resource diversions that effectively restrict low-income households from obtaining NHT housing assistance. The chapter argues that a greater share of NHT's massive financial assets *could* be directed towards serving the housing needs of low-income people if the Trust were organized differently.

The chapter is based on NHT's internal financial records on contributors as of 1980 and 1993 and data on its mortgage recipients as of 1988 and 1995.[1] Much of this analysis is based on a match between contributor and mortgage recipient records using a unique identifier embedded in the data files. Unfortunately, incomplete or inaccurate data allowed us to successfully match only a small portion of the populations in this way.[2] Indeed, it is estimated that in 1995, NHT itself was able to match only 25 per cent of employer contributions with their own payment records over the previous six years (NHT 1995: 42). Even basic information such as the total number of employee contributors or the precise number of mortgages granted to date is unknown.[3]

The chapter will begin by briefly outlining the housing finance problem in Jamaica before providing an overview of the NHT. It will then go on to analyse the socio-economic characteristics of the contributors and mortgage recipients who benefit from the NHT's credit subsidy programme. The chapter will conclude by focusing on initiatives which would ensure a more equitable distribution of NHT resources.

THE HOUSING FINANCE PROBLEM IN JAMAICA

Jamaica's housing problems begin in the formal private sector. Less than 10 per cent of Jamaicans have sufficient incomes or assets to qualify for a private sector mortgage for the cheapest house being constructed (Jones *et al.* 1987, Gleaner 1991).[4] That such a small proportion of Jamaicans can afford a home may seem remarkable but some figures should make the nature and extent of the private and public sector crisis clearer. First, in spite of a relatively high per capita income (US$ 2,532 in 1995), incomes are unevenly distributed between the rich and poor with 41 per cent of employees in Jamaica having an annual salary of US$ 930.[5] Moreover, the income disparity between executives and support staff in the public sector is almost twice that of the private sector, and is getting worse as the salaries of people occupying upper income jobs 'have been increasing very fast in real terms' (Witter 1996: 35). These trends suggest that the high-level public officials who are orchestrating the belt-tightening required under the neo-liberal transformation of the Jamaican economy are themselves reaping considerable material benefits from it.

Second, construction material costs have outpaced inflation at a time when real wages have been declining. Given the problems of exorbitant costs and the low output of the formal sector, most rural homes and almost half of all homes on the island have been built by the occupants themselves (Witter 1996: 31; Eyre 1997). Third, and perhaps most significantly, although public sector housing such as that provided through the NHT, is largely affordable to low-income groups, its distribution is biased towards richer households. For example, a self-help scheme recently offered by the NHT (priced at US$ 12,660) requires a weekly income of just US$ 43 which, according to the NHT's own affordability formula, would be affordable to about 90 per cent of contributors.[6] Indeed of the 15,000 applicants for this scheme 40 per cent had weekly incomes of US$ 56 or less, but only 15 per cent of these applicants were allotted units.

Women, and especially those who are heads of households, have been particularly disadvantaged by the shortfall in housing provision. It is estimated that 37 per cent of all Jamaican households and 45 per cent in Kingston, where housing costs are the highest and self-help housing options are limited due to the scarcity of developed land, are headed by women (Klak and Hey 1992). Fortunately, for these women, Jamaica has an especially rich tradition of people incrementally improving and expanding their houses over time. The 'partner' system, which is an informal savings and loan arrangement among family and friends, and which is almost always headed by women, provides an important source of finance for this purpose. Alternative initiatives, such as the Women's

Construction Collective (WCC), have trained more than 200 women in construction skills since it was established in 1983 to enable them to solve their own shelter problems (WCC 1989). Such creative, self-help strategies have contributed to Jamaica having an unusually high housing stock value compared to the income levels of the occupants (Eyre 1997). The particular housing needs of women, however, have not been central to the state's policy responses and no NHT programme is devoted to meeting women's housing needs as we will see in the following section.

THE NATIONAL HOUSING TRUST: AN OVERVIEW

The NHT was born as a central component of democratic socialism under the PNP government of the 1970s and was seen as 'the most far-reaching and fundamental policy of social change attempted by the Manley government.' (Hope 1976: 10A). The 1976 NHT Act that created the agency pledged that 'initially, low-income contributors are to be given priority consideration'. A second, and seemingly contradictory, expression of NHT's mandate was that housing assistance would be divided equally among income groups.[7] Thus, allocation criteria were designed so that all income groups contributing to the Trust would have more or less the same proportion of mortgages (Smallman 1977). As contributions to the NHT are based on income level, equality could only be achieved by extracting 'social surplus' from the middle-class and above, and redistributing it to the poor (Gilfillian 1978).

The NHT obtains the bulk of its funds from a 2 per cent compulsory savings withdrawal from the payroll of 550,000 formal sector employees and a corresponding 3 per cent tax on employers on the same payroll. This combination of employee and employer contributions amounts to the capture of two and a half weeks' wages per formal sector employee per annum. In the face of currency devaluations and deteriorating terms of trade, the NHT has been fortunate to have a relatively secure and consistent source of funds over time. Employees are entitled to a refund of their contributions plus 3 per cent annual interest on the sum seven years after they have been deducted from the pay-cheque (Gleaner 1975).[8] However, with inflation at an average of 25 per cent per annum (1976–96) the real value of refunds is very small (Planning Institute of Jamaica, various years). Refunds can be and are often seen as political tokens with many contributors never bothering to collect them. The NHT, therefore, benefits from the low-cost of funds (about 1.2 per cent per annum) which is subsidised by the entire formal sector work force.

The political economy and NHT finance

From a position of possessing a reliable and cheap source of funds, the NHT is able to perform four main material roles in Jamaica's political economy. These roles illustrate the various ways in which NHT funds are diverted from being returned to the contributor population in the form of mortgages. First, with 646 employees, the Trust is a significant source of (primarily) white collar jobs. These jobs, however, are provided at a considerable cost. For example, in the 1994–5 fiscal year, this cost amounted to approximately US$ 11 million or 24 per cent of the value of contributions (NHT 1995). These operating costs as a share of appropriated resources are at least twice as high as those in private sector housing institutions and higher than comparable state agencies in other countries (Peterson and Klak 1990, Klak 1993). By halving its operating costs, and thereby putting itself in line with the building societies, the NHT could finance 436 studio units per annum (these consist of a kitchen, bathroom, and one large additional room).

The NHT, therefore, is a major and largely inefficient consumer of its own resources. One measure of the rate of internal consumption is the 'point spread', or the interest rate difference between funds borrowed and funds lent, which the NHT estimates to be 12 per cent. This spread is obtained from comparing the real cost of funds to the 9 per cent that the NHT earns on an average mortgage and what it receives from investments in government securities and property funds in the hotel and tourism sector. The point spread has resulted in an accumulation of the Trust's total assets by 8.5 per cent per annum in real terms between 1976–95. In terms of 1995 assets, this accumulation would be enough to finance the construction of an additional 1,674 studio units.

A second role which the NHT has performed over recent years has been to provide an increasing share of funds as a form of indirect taxation in order to fulfil the reserve requirements of structural adjustment. These mandated reserves, or 'Ministry of Finance Targets' as they are known locally, are the amounts by which short-term investments (usually in revolving securities such as Treasury Bills) need to grow for the government to meet its liquidity requirements. The US$ 31.4 million held in reserve by the Ministry of Finance in fiscal 1995-6 would have been sufficient to finance the construction of 2,485 studio units. Consequently, a significant proportion of the NHT's cash inflow from contributions and investments is not being used for housing.

Third, the NHT uses its assets to provide interim finance to construction firms building both NHT and non-NHT housing. This is done to increase formal sector production which has been otherwise restricted by high interest rates of between 30 and 50 per cent per annum during the 1990s (Witter 1996). Although the precise financial value and number of non-NHT housing units that have been financed in this way is unknown even to NHT, one result of this involvement in interim financing has been to decrease the proportion of low-income housing. In addition, total housing expenditure has decreased in real terms for most years since 1978: 1995 expenditure was 12.4 per cent lower in real terms than in 1978.

The Trust's fourth, and perhaps most significant, role is to provide contributors with mortgages at highly subsidized rates of interest. With inflation averaging 25 per cent per annum and a mean mortgage interest rate of 9 per cent (which is not indexed to correct for inflation) this vast interest subsidy is in practice paid for by the full population of contributors. We now turn our attention to a comparison of the NHT's contributors and mortgage recipients to reveal in more depth who pays for and who benefits from the credit subsidy programme.

THE NHT'S CREDIT SUBSIDY PROGRAMME: WHO PAYS AND WHO BENEFITS?

Our starting point for examining the socio-economic characteristics of NHT mortgage recipients is to look at the number and size of mortgages granted over time. The annual variations in the number and size of mortgages is attributable to a host of factors. These include policy shifts within the NHT, the number of qualifying applicants and the financial performance of the Trust in the preceding year. In addition, external factors such as the vision for the Trust held by the political party in power, the current economic environment and the relative availability of financial capital and land also have an impact on the number and size of mortgages available.

As shown in Figure 4.1, the trend in the delivery of mortgages over the last two decades has been modestly upward. Most of the increase occurred in 1995–6 when mortgage allocations were double the average level of previous years. This recent surge reflects a policy shift towards providing mortgages for alternative housing programmes such as the provision of studio units rather than multi-bedroom completed units.

Figure 4.1 Distribution of mortgages by the year the loan was committed
Source: NHT's 1995 Mortgage Master File.

The doubling of the annual number of mortgages might be regarded as a socially progressive trend as, theoretically, the greater the number of contributors to obtain housing assistance and enjoy the interest subsidies so the wider the distribution of benefits. However, more mortgages do not necessarily mean that more low-income people obtain them, although they are more likely to qualify for the mortgages made for the incremental housing programmes. An examination of the allocation of NHT mortgages reveals that even within a contributor population that excludes lower-income informal sector workers (especially women) mortgage recipients have a relatively high income (Table 4.1). For example, although less than 2 per cent of the contributors are paid over US$ 29,000 per annum, they have succeeded in securing 34 per cent of the NHT's mortgages. Conversely, the 71 per cent of contributors who are paid less than US$ 4,400 per annum have obtained less than 48 per cent of total mortgages. As shown in Table 4.1, the median income for mortgage recipients is almost twice the national income and more than twice that of all contributors. As of 1988, the probability of obtaining a mortgage was seven times higher for a contributor in the ninth income decile than for one in the bottom third of contributors. Contrary to the NHT's mission statements, then, low-income contributors have not been the primary beneficiaries of subsidized mortgages.

From an examination of the gender of mortgage recipients, it appears that even within the current NHT system, the share of mortgages made to women could be higher if there was a policy to target women. The high proportion of female-headed households in the country is one reason for treating women as a special group. Another is the predominance of women in the long queues of people seeking applications and other information from the NHT which attests that women spend more time than men pursuing government housing assistance (Klak and Hey 1992). However, in

Table 4.1 Distribution of annual income of mortgage recipients and contributors

Income groups (US$)	Number of contributors	Number of mortgage recipients
less than 1,612	58,878 (40)	1,775 (14)
1,613–2,912	28,618 (20)	2,417 (19)
2,913–4,368	16,612 (11)	1,834 (15)
4,369–7,280	19,738 (14)	1,182 (10)
7,281–29,224	20,043 (14)	904 (7)
29,225–58,500	1,603 (1)	1,274 (10)
58,501–87,800	253 (–)	1,285 (11)
Over 87,800	284 (–)	1,729 (14)
Total cases	146,029 (100)	12,400 (100)

Source: NHT's 1995 Mortgage Master File.
Note: Figures in parenthesis are column percentages.

spite of the fact that no NHT programme specifically targets women, it has to be recognized that women in Jamaica acquire a reasonable proportion of NHT mortgages by international standards. For example, the percentage of women among all mortgage recipients rose from 40.5 per cent in 1988 to 45 per cent by the end of 1995. Indeed, women contributors have a slightly higher probability of receiving a mortgage than their male counterparts and are able to obtain comparable-sized mortgages with incomes that average only 75 per cent of those of men.

That women have received a high share of mortgages is mostly attributable to the actions of women themselves rather than the NHT. Part of the explanation is that women now make up 47 per cent of the labour force and their demand for housing has increased along with the more general economic crisis affecting Jamaica. In addition, the fact that women comprised 56 per cent of all state workers as of a few years ago means that they have used public sector employment as a vehicle to access housing assistance. Biases in the allocation of NHT mortgages mean that even though state employees represented only 9 per cent of all workers during the late 1980s, they managed to secure 58 per cent of NHT mortgages (Klak and Hey 1992).

An examination of the contributor population reveals further biases. Although the Trust's mandate is to provide benefits based on need, an individual must be a current contributor to be eligible to apply for a mortgage. The contributor population is largely comprised of people working in the formal sector. As an estimated one-half of the working population is employed in the micro- or informal sector this means that many Jamaicans are unlikely to be NHT contributors as their incomes are too low to make regular payroll contributions. This is shown by the fact that self-employed persons constitute only about 5 per cent of the NHT's contributor population.

Likewise, low-income women are under-represented in the contributor population. The NHT's apparent favouring of women as mortgage recipients does not overcome the effects of the overall biases that prevent women, particularly low-income women, from entering the contributor population. This is partly attributable to, first, the fact that Jamaican women, like their counterparts throughout Latin America and the Caribbean, represent a disproportionate share of the self-employed and micro-enterprise labour force (Rakowski 1994). Such informal employment carries none of the documentation and social legitimacy of formal-sector wage statements, and therefore informal workers lack the essential ingredients of a successful mortgage application. Second, unemployment among women has been relatively high. By the end of 1993, for example, while male unemployment stood at 11.1 per cent, unemployment among women was nearly 21.5 per cent (Statistical Institute of Jamaica 1994). The high unemployment level among women is partly due to the neo-liberal transformation in the 1980s which has resulted in massive employment retrenchment in the public sector. Women's employment has often been the first to go.

From the above analysis it would appear that the NHT neither serves the low-income population with subsidized mortgages, nor draws this population in as contributors. Those households most in need of assistance therefore are excluded from the NHT credit subsidy programme. The following section identifies the reasons behind this.

Factors accounting for the distribution of NHT resources

The regime effect

To what extent has the income level of the average mortgage recipient changed over time, and in particular, with respect to the political party in power?[9] Over the NHT's lifetime, the country has experienced political changes as the People's National Party (PNP), in power in the 1970s, has been followed by the Jamaican Labour Party (JLP) in the 1980s and subsequently the PNP again. Although both parties are populist and clientelistic, the JLP is politically right of centre and favours private-sector solutions while the PNP has its roots in left-wing ideology. In the past, the PNP has pursued state economic planning and the progressive redistribution of wealth but in the neo-liberal era, it has shifted markedly towards the right and the position of its political adversary.

The distribution of mortgages by income groups over the years indicates a temporal pattern associated with regime change (Figure 4.2). More low-income people received mortgages in the 1970s and 1990s under the PNP than in the 1980s under the JLP. The probability of acquiring a mortgage was greatest for those persons in the sixth through eighth income decile during the

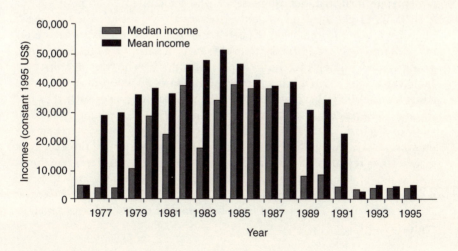

Figure 4.2 Median and mean incomes of mortgagors, 1976–95
Source: NHT's 1995 Mortgage Master File.

1970s. Under the JLP Seaga administration, the NHT's lending favoured the contributors from the two highest-income deciles. Although concerns about the accuracy of income data for the 1990s prohibit us from conducting a similar analysis for the present decade, we are confident that such an analysis would reveal that the income deciles most likely to obtain a mortgage has fallen since the 1980s.

It is clear that the mean and median income of mortgage recipients has decreased most dramatically since 1991. Milverton Reynolds, NHT managing director since 1990 (soon after the PNP returned to power), has made some policy changes such as the diversification of the NHT's housing programme so as to create more lower-cost housing alternatives. These include the provision of serviced plots, studios and two- (rather than three- or four-) bedroom scheme units, and the use of smaller contractors who are able to keep down construction costs. The NHT is also renewing its commitment to provide loans for construction materials while low-income mortgage recipients invest their own labour to keep costs down (this is called the 'Build on Your Own Land' programme) (Planning Institute of Jamaica 1992). Between 1990 and 1995, scheme units were about 33 per cent of all mortgages while home improvement mortgages were 24 per cent and 'Build on Your Own Land' were 31 per cent of total mortgages. These cheaper options attract more lower-income applicants and, of these, a greater proportion qualify for, and have received, mortgages during the recent past.

Unfortunately the lower incomes of the average mortgage recipient in the 1990s are partly attributable to factors other than a re-targeting of housing policy. Two factors are notable. First, the NHT's 1995 mortgage files exclude about half of all mortgage recipients since the 1990s. Among the mortgages not entered in the data set are those for which the NHT combines its funds with those of private lenders and a para-statal institution, the Caribbean Housing Finance Corporation (Klak 1992a). Although impossible to determine definitively, it is likely that these excluded mortgages have gone primarily to higher income groups. Second, for the vast majority of Jamaicans, long-term high inflation, economic recession and structural adjustment policies have seriously eroded purchasing power such that real incomes have declined significantly (Witter 1996). For example, the real value of teachers' salaries at present has less purchasing power than that of a domestic worker in the 1970s. A recent NHT analysis of the professions of beneficiaries reveals that those traditionally considered 'market occupations' such as nursing and teaching now fall within low-income categories. It would seem, therefore, that while in the past the nature of the regime in power was important, the more market-oriented behaviour of the 1990s transcends the regime effect.

Interest-rate policy by income level and mortgage size

Interest rates are a crucial variable in housing policy which affect mortgage affordability and programme costs, as well as the level of redistribution or cross-subsidization among mortgage recipients and the real cost of subsidies. A central component of NHT's mortgage allocation system is the manipulation of interest-rate policy. From its inception, the NHT has claimed that its interest rates increase as a function of mortgage recipient income and it has pursued a 'progressive interest rate' structure with the intent of cross-subsidizing from high- to low-income groups. During this period, interest rates have ranged from 4 to 10 per cent. More recently, in an effort both to expand affordability for low-income groups and to narrow the gap between NHT interest rates and those in the private sector, the Trust has introduced a rate of 2 per cent and 15 per cent for low- and high-income contributors respectively.

Despite the centrality of differential interest rates to NHT policy, the data indicate that their effects on mortgage granting practices have been minor (Table 4.2). Notwithstanding the NHT's rhetoric of a progressive interest-rate policy, over 85 per cent of all mortgages continue to bear an interest rate of 8 or 10 per cent, and therefore show little relationship to income. Those in the lowest income category have been charged a mean interest rate of 8.3 per cent and a median rate of 10 per cent while the highest income group has been charged a mean of 9 per cent and a median of 10 per cent. The highest mean interest rate is actually paid by the fourth to eighth income categories.

If a large range of interest rates is not charged on the mortgages made, then differences in mortgage size offset the effects of progressive interest rates (Peterson and Klak 1990). Any policy of cross-subsidization is, therefore, undermined since larger mortgages will receive larger subsidies and these larger mortgages usually go to higher-income groups. It follows that the correlation between mortgage size and interest subsidy is nearly a perfect one $(r = +0.973)$.

Table 4.2 Mean and median interest rates by annual income of mortgage recipients

Income groups (US$)	Interest rates		Percentage of mortgage recipients
	Mean	Median	
less than 1,612	8.3	10.0	14
1,613–2,912	8.4	10.0	19
2,913–4,368	9.1	10.0	15
4,369–7,280	9.5	10.0	10
7,281–29,224	8.9	10.0	7
29,225–58,500	8.0	8.0	10
58,501–87,800	8.4	8.0	11
Over 87,800	9.0	10.0	14

Source: NHT's 1995 Mortgage Master File.

Subsidies by income level and mortgage size

Interest rate-based subsidies can be measured as a function of the difference between the 'market' rate (i.e. the building society's 19 per cent) and the NHT's mortgage rate. It is important to note that this comparison yields a very conservative estimate of subsidies for several reasons. First, the Jamaican government has held the building society's interest rate well below commercial rates and inflation in recent years. Second, NHT houses usually sell significantly below market prices and increases in home values have outpaced inflation. Third, as of 1988, the average mortgage enjoyed an annual subsidy of about 15 per cent of its value due to payment arrears.[10]

Notwithstanding the above, to approximate the value of interest subsidies we calculate that the mean annual interest rate subsidy for all mortgages is US$ 1,496, the median subsidy is US$ 1,125 (or 44 per cent of the national median income), while the total annual subsidy is US$ 43.5 million. As a measure of opportunity costs, the NHT's annual interest subsidy could finance the mortgages for 3,433 studio units at a price of US$ 12,660.

Which income groups are benefiting the most from these interest subsidies? When interest rates and mortgage sizes are combined one finds that subsidies increase with income. Mortgage recipients in the lowest NHT income categories obtain the smallest interest subsidy while those in the highest income category reap the largest subsidy. However, there is no consistent pattern across all income groups and subsidies are widely scattered with respect to income (Figure 4.3). In 1988, for example, although the correlation between income and interest subsidy is positive it is very small (0.04). This may seem counter-intuitive but can be explained by the fact that mortgage recipients have only rarely obtained the maximum NHT mortgage for which their incomes qualify. Furthermore most mortgages are extended for scheme housing for which a wide range of income groups may apply. The price for these units is fixed by the NHT as a function of construction costs and mark-up charges. Many of those selected have income levels well above the minimum necessary.

The randomness of interest subsidies with respect to income makes them ineffective as a (progressive or regressive) redistributive policy. In effect, interest subsidies are redistributed from the many contributors (more than 90 per cent of whom work in the private sector) to the few mortgage recipients (the majority of whom are state employees). The correlation between income and interest subsidy has increased since 1988 because house prices have increased faster than most incomes. This has forced more people to take a mortgage that is closer to the maximum that they can afford.

As Figure 4.3 suggests, the high-income group enjoys large subsidies in absolute monetary terms, but small subsidies as a percentage of income. The average interest subsidy for someone earning US$ 87,000 is around US$ 1,800 per annum, or approximately 2 per cent of income. In contrast, the lowest

70

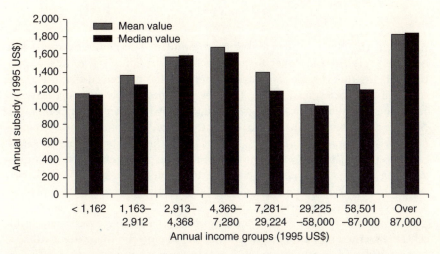

Figure 4.3 Distribution of mean and median interest subsidies by income groups
Source: NHT's 1995 Mortgage Master File.

income group enjoys smaller subsidies in absolute terms which are relatively large when compared to their income levels. For a mortgage recipient earning just US$ 1,100 per annum, the average interest subsidy is approximately US$ 1,100 per annum. Although smaller in absolute terms, the NHT's interest subsidy effectively doubles the income of the relatively small share of low-income contributors fortunate enough to obtain a mortgage.

When comparing contributions to the Trust to mortgage amounts for each income level, an element of social progression within the contributor population emerges. The largest re-allocation of housing funds has gone to the bottom third of wage earners. As contributions are a function of income, and income is very unevenly distributed in Jamaica, low-income people contribute relatively little to the Trust. In contrast, the vast majority of the housing that NHT has made available has a high selling price compared to income, and especially that of the low-income households.

The problem for low-income people, however, is that they have been granted relatively few mortgages even though these vastly offset the tiny payroll deductions of the low-income group. A contributor in the bottom one-third of the income distribution is highly unlikely to receive a mortgage. In other words, among those whose need for housing assistance is greatest, NHT has provided only a few, exceptional mortgages. This reflects the fundamental contradiction in NHT's method of gathering funds and allocating housing assistance. Low-income people are unable to contribute to the housing fund in sufficient quantities to finance their own housing and they do not have adequate incomes to afford the monthly mortgage for completed housing units.

CONCLUSION

Although the NHT has been able to accumulate a huge asset base, it has delivered very little low-income housing. Against such criticism in the past, the NHT's senior director for corporate services has responded that the Trust cannot be held responsible for the low incomes of the vast majority of Jamaicans (George 1996). Of course it cannot. However, as a public agency, the NHT is responsible for living up to its original and ongoing mandate to prioritize low-income housing needs and for using the mandatory contributions of workers for this expressed purpose. The tens of thousands of Jamaicans with low incomes who lack adequate housing should be provided with shelter assistance at a price appropriate to their purchasing power at a scale appropriate to NHT's vast resource base.

Presently most of the NHT's funds go to meet its own payroll, help fulfil central government reserve requirements, provide interim financing of higher-income and even non-NHT housing construction, and finance the mortgages of a relatively few and mostly higher-income contributors. In this chapter, we have measured the scale of resource diversion towards uses other than low-income housing and have noted how many relatively affordable and incremental studio units could be financed with those funds. First, the US$ 31.4 million diverted to meet reserve requirements in the fiscal year 1995–6 is enough to finance 2,485 studio units. Second, by halving its operating costs, and thereby putting itself in line with the building societies, the NHT could finance another 436 studio units per annum. Third, the 8.5 per cent growth in the NHT's real assets for 1995 is enough to finance 1,674 studio units. Fourth, the NHT's annual interest subsidies could finance the mortgages for 3,433 studio units. When combined, these diversions are equivalent to the cost of constructing 8,028 units per annum. Such provision would be more than double the NHT's yearly output of housing mortgages and bring it closer to meeting the 15,000 new units that Jamaica needs each year.

A very important change in NHT policy would be for a greater share of assets to be channelled to the Trust's original purpose of prioritising the housing needs of the poor. One way in which this could be done would be if the NHT were to use its funds to help community-based organizations, such as NGOs, deliver housing finance to low-income groups. By doing so, NHT funds would be used in projects which have a direct impact on low-income housing conditions. Indeed, the NHT periodically makes monetary donations to some NGO activities in the Kingston metropolitan area such as the Mustard Seed Organization. This NGO-NHT joint venture has resulted in the construction of 80 low-income housing units in a deteriorated area of downtown Kingston and has employed local residents to keep costs down. Residents who are members of credit unions, building societies and/or the NHT can apply for housing mortgages. This joint initiative illustrates that combining NHT low-interest funds with alternative finance makes these mortgages more affordable to low-income men and women. Such programmes could be made sustainable in the long run by indexing the money invested to inflation so that its value is maintained. Moreover, small mortgages could be extended to both NGOs and households, with the former being responsible for managing mortgage portfolios.

At present, the NHT's inefficiencies are apparent to anyone who closely examines its pattern of finance use. The NHT should be clearer as to what its contribution to the housing system is, which socio-economic groups benefit and how much it spends internally. On this point at least, we are in accord with neo-liberals who call for greater transparency in the workings and rules of state organizations like the NHT. What is clear is that there needs to be a major overhaul of the NHT's spending allocations.

NOTES

1 It is important to point out that computer records for NHT mortgage recipients are several months out of date, and those of contributors are several years behind. While we have tried to use the most recent data, in some cases concerns about the reliability of information has led us to rely on earlier and more trustworthy figures.

2 We would have liked to obtain more complete information about how many low-income contributors who apply for mortgages are selected and how this has varied over time. Unfortunately, the inconsistent way in which such data have been recorded severely limits analysis and interpretation. Thus, we have had little choice but to focus on the complete records of contributors and mortgage recipients. It is important to bear in mind, however, that many lower-income contributors are excluded during the application process.

3 Such data management problems certainly do not inspire confidence in the organizational capacity of the NHT.

4 Forty-three per cent of the cost of a new house goes to profit, interest and taxes. Private-sector mortgage interest rates begin at 19 per cent per annum (Witter 1996, Eyre 1997).

5 All monetary values have been converted to 1995 US$ (US$ 1 = $JAM 35.54). For lack of a better alternative, the rate of inflation was used to correct and standardize Jamaican dollar values over time. Thus, the real value of funds such as housing investments, mortgage values and incomes for years prior to 1995 were approximated by adjusting them by the annual inflation rates for all years separating them. The accuracy of this conversion is open to question. Certainly, income levels for the vast majority of Jamaicans have not kept up with inflation, which means that the estimates of the value of 1995 real incomes are increasingly exaggerated as one goes back in time. The difficulty in approximating real values over a period of high, long-term inflation serves to highlight the challenges faced by NHT in its quest to provide effective housing solutions.

6 While NHT claims to grant mortgages to any contributor who can afford them, what NHT income groups can actually afford with a NHT mortgage is controversial. Earlier analysis by Klak (1992b) estimated that the median mortgage recipient paid just 7 per cent of income towards her/his mortgage. Contrary to conventional wisdom, the correlation between the number of months a mortgage is in arrears and the mortgage/income ratio is actually negative ($r = -0.16$). That research also found that while income and the number of months mortgage payments in arrears is modestly negatively correlated ($r = +0.10$), there is virtually no statistical relation between

income and monetary value of arrears (r = +0.05) (Peterson and Klak 1990: 20, Klak 1992b).

7 These two mandates reveal that the NHT attempts to satisfy the needs of a variety of interest groups whose priorities are in conflict. Thus, although the NHT theoretically aims to allocate housing finance in a manner more 'socially progressive' than the private sector, in practice, it performs much like the private sector.

8 From 2001, the NHT will begin to refund employer contributions made before 1980 (NHT 1995).

9 The mean and median do not always co-vary and in particular the mean is often much larger than the median. When this happens, as in 1977–9 for example, the distribution of mortgage recipients is skewed towards those with very high incomes. Still, the temporal pattern over the years is fairly consistent whether central tendency is measured by the mean or median.

10 As of 1988, 94 per cent of mortgage recipients were in arrears, with a mean period spent in arrears of 11.8 months (Klak 1992b). In 1997, the value of arrears was put at US$ 5.4 million even though all arrears on mortgages before 1993 had been rescheduled.

5

FROM COMMERCIAL BANKING SYSTEMS TO NON-COMMERCIAL BANKING SYSTEMS IN MEXICO

William J. Siembieda and Eduardo López Moreno
California Polytechnic State University, USA,
and Universidad de Guadalajara, Mexico

INTRODUCTION

Three factors are important in understanding the status of the housing crisis in Mexico. First, there is a housing deficit that exceeds 3 million units and is increasing at an annual rate of over 200,000 units (SEDESOL 1996). In order to address this housing deficit a number of ambitious government programmes have been formulated over the past twenty-five years, but supply has only ever been able to address about 20 per cent of the total annual demand (Siembieda and López Moreno 1997, Pezzoli 1995, Ward 1991). Second, the purchasing power of wages has declined to approximately 40 per cent of 1985 values, presenting low-income households with problems of affordability (González de la Rocha 1994, Ward *et al.* 1993). Third, the housing finance system is inadequate either to address the demands for finance from households or to motivate a more efficient supply from developers (Zearley 1993, Lea 1996).

While these problems are acknowledged by the Mexican government and have prompted reform of the public and private commercial housing finance system on numerous occasions since the mid 1980s, this chapter argues that such reforms have not had the desired effects. An examination of the deficiencies in the present system of commercial and public bank trusts reveals significant constraints on the government's ability to raise real wages and to incur debt from housing subsidies. It seems unlikely that the further continuation of a policy to readjust the present housing finance system, albeit in increasingly market-oriented forms, can ever meet the country's real needs. Furthermore, a series of events in the finance sector specifically as well as the problems of macro-economic management, mean that the people's trust in the central government or the commercial banking system to solve the housing crisis is low. It would seem to make policy sense, therefore, to promote housing finance delivery systems in ways that can mobilize support and savings through, for example, the growing number of non-commercial bank experiences that are starting to emerge at the local level. We argue that the government needs to give greater emphasis to the promotion and development of financial programmes that are locally based but nationally supported. Future policy in the housing sector must be broader, decentralized, and culturally relevant, particularly to the working class or *popular* sector in order to promote flexibility and adaptation to local conditions.

THE FORMAL HOUSING FINANCE SYSTEM

Mexico's support for housing finance began over fifty years ago. In 1947, the Fondo de Casas Baratas was set up through the National Bank for Public Works (BANOSPSA) and in 1954 the Instituto Nacional de Vivienda (INV) was established to assist formal sector workers with cheap finance. Separate institutions were formed to subsidize shelter for worker groups in the oil, electricity supply, teaching and railway sectors (Garza and Schteingart 1978). The creation of the Programa Financiera de Vivienda in 1964 established housing and housing finance as part of national public policy (Garza and Schteingart 1978, Ward 1991). The Programa Financiera de Vivienda formalized the principle of the government providing the capital for the construction of 'public' or 'social interest' housing. Perhaps the most successful agent to conduct this task has been the Fondo de Operación y Financiamiento Bancario de la Vivienda (FOVI).[1] Established in 1963 as a trust of the Banco de México, FOVI had the objective of channelling mandatory investments by commercial banks for the construction of 'social interest' housing at below-market interest rates to households with incomes 2.5–5 times the minimum salary, about US$ 214–428 per month (Ward 1991). Originally, the banks were required to provide 3 per cent of their lendable assets for FOVI credits, a proportion that was raised to 6 per cent in 1984 as part of the National Housing Finance Programme.[2]

Although, in the past, the Mexican government has built housing directly, the quantities have always been small and limited to special projects. A more accurate description of Mexico's 'public housing' agencies is as trusts (*fondos*) that provide low-cost capital rather than housing. Whereas FOVI operates as a public–private partnership between the commercial banks and private developers, Mexico's three principal public housing agencies have traditionally supplied capital to corporate agents such as labour unions or civic groups at below-market rates on condition that the final housing be affordable to particular social groups. The two largest agencies, both established in 1972, are the Instituto Nacional del Fondo de Vivienda para los Trabajadores (INFONAVIT) and the Fondo de Vivienda del Instituto de Seguridad y de Servicios para los Trabajadores del Estado (FOVISSSTE). These two trusts are financed from mandated contributions from formal sector workers and employers in the private sector (INFONAVIT) and the public sector (FOVISSSTE) equivalent to 5 per cent of the wage bill (INFONAVIT 1987).

The third housing agency is the Fondo Nacional de Habitaciones Populares (FONHAPO) which was set up in 1982 to replace INDECO, itself established in 1971 as a replacement for the INV (Ortíz 1995, Ward 1991). FONHAPO is supported (as previously was INDECO) by federal budget transfers from the national public works bank (BANOBRAS), but operates in a more decentralized way than INFONAVIT or FOVISSSTE through a series of state-level housing institutes. Its innovative finance system aims to target the low-income population, mainly non-salaried workers, with incomes less than two-and-a-half times the minimum wage through provision of collective loans (Coulomb 1990, Ortíz 1995). In order to retain a notion of affordability, FONHAPO encourages shelter solutions in the form of sites-and-services projects and progressive construction with important roles conducted by community participation (Mitlin 1997, Ortíz 1995).[3] Furthermore, unlike INFONAVIT which is closely tied to the ruling party through the labour unions, FONHAPO has been able to provide finance to community organizations that have no direct clientelist relations with the ruling political party (Ziccardi and Mier y Terán 1990).[4]

Difficulties in the Mexican housing finance system

From 1983 to 1988, the Mexican housing finance system funded the construction or substantial improvement of approximately 223,000 units each year with an investment in excess of US$ 2.8 billion. During this period commercial banks had only marginal participation. But from 1989 to 1994 the finance system raised its level of funding to 441,000 units a year with an investment in excess of US$ 5 billion (Joint Center for Housing Studies 1997). During this period the private banks were providing about 50 per cent of housing investment with an additional 7-10 per cent through FOVI, the payroll funds were contributing about 20–30 per cent and the other housing agencies about 12 per cent. Despite contributing the dominant share of investment, the commercial banks and FOVI were financing less than one-third of all new house starts. Moreover, the housing finance system was not addressing the needs of 55 per cent of the population that was forced to self-finance their housing through the popular sector (Ward 1991). By the end of the 1980s there was also a diversification of finance by housing type: 75 per cent of finished housing was for social interest projects; 13 per cent was progressive housing and a further 8 per cent was sites-and-services; and upper income housing received 4 per cent (Joint Center for Housing Studies 1997).

Although this level of finance provision can be viewed as a success, it can be argued that Mexico's housing finance system contained three important, but not mutually exclusive, structural weaknesses. First, the system was too

segmented with the result that it has produced many agencies and programmes that have played towards discrete political constituencies and have
not prioritized national coverage. In particular, observers have noted that the
system affords too much influence over the flow of funds to special interest
groups such as the trade unions with the result that expenditure is often inappropriate to need (Pezzoli 1995).[5] Second, the system has been highly
dependent upon external capital and affords very little credence to the possibility of generating finance flows from internal capital sources.

The third weakness is that the housing finance system is too centralized
with the federal agencies dominating local ones and national commercial
banks crowding out smaller regional credit programmes. Critics have argued
that this centralized system has been driven by an assessment of national
requirements rather than the needs of households in particular locations
(Comité Preparatorio de México 1996). In order to maintain a centralized
system it is argued that planners have sacrificed equity as nationally focused
schemes have little capacity to adapt to local conditions. Contrary to the seg-
mentation view, therefore, critics argue that the over-centralization of the
housing finance system has limited the development and diversity of financial
systems and created oligopolistic markets for both construction and mortgage
finance.

Partly related to these structural weaknesses, the housing finance system has
been highly inefficient and incurred high operating costs. The commercial
bank sector, for example, has accounted for about 21 per cent of the total units
to the housing stock, but has consumed nearly one-half of all housing funds
(SEDUE 1992). The payroll trusts (INFONAVIT, FOVISSSTE) have faired
little better, providing one-third of total formal housing production but
absorbing about 44 per cent of total housing investment (SEDUE 1992). In
delivering these units, the system has operated at a financial cost that is significantly higher, by perhaps as much as 50 per cent, than that of non-financed
private developers. By contrast, FONHAPO has managed to deliver housing
at reasonable cost relative to low-income affordability thresholds and at costs
commensurate with the best performing private developers.[6] Thus, with about
9 per cent of the federal housing budget FONHAPO has contributed a disproportionate number of housing starts, usually about 15 per cent of
public-sector assisted supply (Coulomb 1990, Pezzoli 1995). This is the best
ratio of investment to production of all the funding strategies in the Mexican
housing finance system.[7]

There is almost unanimity of opinion that the housing finance system has
not met the needs of low-income households in sufficient numbers, but has
channelled finance to a narrow sector of the population. FOVI, for example, has
financed projects with housing valued at 100–220 minimum annual salaries
and, with commissions and legal fees, deposit requirements (usually 20 per
cent) and service charges, only households with high and stable incomes more
than five times the minimum wage are eligible. This income bias has been

extended by a bias towards home-ownership of finished units at the expense of rental or progressive housing: in the state of Jalisco from 1980 to 1990, 87 per cent of housing investment was assigned to finished housing. As a consequence, nationally, only 11.9 per cent of the population, the so-called creditworthy, are able to gain access to the finance available and, as real wages have declined during the 1980s, there have been fewer qualified applicants over time.

Recent reforms to the housing finance system

That these criticisms and weaknesses, many of which were made with reference to the system's performance during the 1980s, could also be applied in the mid-1990s is all the more reproachable because of the seemingly ceaseless reforms of the housing finance system over the past decade. These reforms have concentrated upon removing restrictions and raising the volume of lending, making the system of allocation more transparent, and addressing cost recovery and equity issues.

A number of steps have been taken to improve transparency. In the case of FONHAPO this process was well under way by 1986 when its guidelines were changed to require public tenders for house construction and infrastructure installation. This measure was designed to boost an ailing construction sector with the result that, in just one year, private sector participation rose from 9.2 per cent to 71.2 per cent. In 1989, both INFONAVIT and FOVI modified their finance allocation procedures, which had long been associated with favouritism, through a programme called SUBASTA. Under the new programme, funds were allocated through an auction where promoters would bid and adjust profits and project design according to the cost of funds. Although only partially successful as a means of delivering more housing, SUBASTA is credited with having distributed funds on a less partisan basis.

In 1993, President Salinas de Gortari reorganized the payroll trusts from a system of direct project finance into a general housing finance system whereby finance would be made available to any institution (including civic organizations) that met the financial limits and building conditions. Whereas in the past the allocation of housing units to households was conducted randomly (in principle) or according to political affiliation (in practice), the reform established a ranking system based upon transparent criteria such as the number of children, years in the system and housing need. A worker could visit any local INFONAVIT office to make an application and receive a ranking classification. The aim of the new system was to diminish the role of developers and unions in the distribution of finance, as had been the traditional practice, and to open up the market to the private sector to offer housing designed to meet local needs. In effect, INFONAVIT's operating rationale was shifted from being a workers' trust fund to a more conventional form of mortgage bank.

The government has also attempted to establish new or to strengthen existing financial instruments. Mexico has gone to considerable lengths to develop an advanced financial services sector including the creation of a secondary mortgage market and mechanisms to allow government agencies to float bonds on the stock exchange (*bursatilización*) (Jones *et al.* 1993). In 1984, FOVI and Banamex, Mexico's largest commercial bank, introduced the dual-index mortgage (DIM) to counteract the effects of inflation rates, which reached 160 per cent in 1987. The DIM separates the payment rate from the amortization rate so that the borrower is protected from extreme changes in monthly payments by indexing repayments to the minimum wage and the amortization rate to either the national short-term Treasury Bond (Certificados Tesoreros, CETES) or the bank rate. The difference between the repayment and amortization rate is then added to the credit balance and the repayment period extended accordingly (Buckley 1996, Lea 1996).

Unfortunately, the operation of DIMs has not resolved the problem of affordability. Although, in principle, a DIM was supposed to be repaid in thirty years, the commercial banks have tended to add at least five margin points to the cost of funds, making repayment difficult over such a timescale.[8] Furthermore, with the austerity measures brought in following the 1994 peso devaluation (from 3.3 pesos to the US dollar in October 1994 to 8.2 in October 1997), the cost of funds has risen faster than wages so that borrowers have owed more than the original loan (negative amortization). In 1995, the average interest rate was more than 62 per cent and the commercial banks rushed to restructure their portfolio as the number of mortgage defaults rose over 250 per cent between 1994 and 1995. The freeze on new mortgages and tighter qualification criteria reduced the number of potential buyers and caused housing production to drop by 53 per cent and demand to dry up, leaving developers with an estimated unsold stock of 60,000 finished houses (Cámara Nacional de la Industria de la Construcción 1996). Consequently, in 1995, the government introduced the Unidad de Inversión (UDI), a new unit of account designed to maintain a constant real value by linking rates to the national consumer price index which is adjusted daily. As with DIMs, a system of dual payment protects the borrower and allows for faster repayment if wages or real property values rise. The advantage of the UDI over the DIM is that it allows the banks to sell their loan portfolio to a government trust and thereby lower their debt exposure.

Under pressure from the World Bank, the Mexican government has also sought to adopt full-cost recovery in housing finance programmes and to reduce subsidies.[9] There was widespread concern, for example, that the system of contract lending operated by FOVI was penalizing the commercial bank sector and was costing US$ 300–500 million per annum by the late 1980s (Lea 1996). Thus, while FOVI itself has been well managed with positive net incomes it was believed that it could go further to reduce subsidies. A recent US$ 900 million loan from the World Bank will support a new strategy of using an initial subsidy tied to a market-based rate for the loan balance (in 1997 the real interest rate was 5 per cent). In this way the consumer will reduce the size of the loan while paying the remainder under a bracketed variable-interest system. It is hoped that this mechanism will provide incentives for rapid repayment and fewer defaults while simultaneously reducing the subsidy portion of the loan.[10] If successful, the expectation is that FOVI will be able to reach households earning between three to four times the minimum wage and open FOVI to a potential market of around 40 per cent of the active labour force instead of the 11.9 per cent that it covers at present.

In the case of INFONAVIT, between 1972 and 1987, it is estimated that borrowers were receiving a 30 per cent subsidy on loans through negative real rates of interest and gratuitous services. According to a study by the Banco de México, under an assumed 100 per cent average inflation rate INFONAVIT was capable of recovering only 14 per cent of its total loans (Banco de México 1987). In 1987, therefore, a new system linked repayments to a predetermined number of multiples of the minimum wage with the result that cost recovery has risen to 62 per cent. The transition towards greater cost recovery, however, has produced a contradiction in the operation rationale for FONHAPO. In 1987, FONHAPO adopted a new financial scheme to allow it to recover half of the loans given to an individual project. While financially successful, cost recovery meant that FONHAPO had to treat its client group, the poorest sector in Mexico, as if it were serving an economic class capable of making long-term repayment. Under this new thinking the government sought to recover investment without giving credence to the social and political pressures that existed within the country. Continued financial pressure to reduce public subsidies has lowered FONHAPO production each of the last three years and it no longer functions as a significant actor in the housing sector.

With a continuing economic crisis in Mexico, can we expect the government and commercial bank sectors to contribute a more significant proportion of national housing production over the next five years? With respect to INFONAVIT and FOVISSSTE, there is likely to be a continuation of the structural shift away from a dependence on fixed percentage payroll funds in order to maintain production levels. The economic crisis, if it continues, will mean that the productive sector will not generate new employment and structural readjustment will require the government to reduce the workforce such that new payroll contributions will stagnate or decline. This means that the public housing sector may be obliged to lower the number or the amount of loans distributed per year. Perhaps in anticipation of such events, INFONAVIT has begun to liberalize access to funds through co-financing arrangements with commercial banks, whereby worker accounts can be used as guarantees for bank loans (INFONAVIT 1996). Since 1997 there have also been moves to raise the capitalization of FOVI and INFONAVIT by allowing national pension retirement trusts to place monies in a new fund, the Administradora de Fondos para el Retiro (AFORES) which then purchases blocks of government-issued housing credits and thus to recapitalize the trusts. The future role of the commercial sector is more difficult to predict since the restructuring of the banking system is likely to continue as the leading banks are no longer protected from foreign acquisition. Nevertheless, we predict a decline in formally financed housing production in the short term.

THE NON-COMMERCIAL BANK SECTOR

Successive economic crises, the financial failure of some banks in combination with the aversion of working-class people to participate with banks and the lack of financial products that fit the needs of people with low incomes, have provided the operating space from which non-commercial bank institutions have begun to arise. This process has been supported by the National Housing Programme (Programa de Vivienda 1995–2000) that calls for more accessible financial schemes, including ways to link the potential creditworthiness of working people with finance and to create a system for savings accumulation in non-commercial bank systems (SEDESOL 1996). The programme points out the need to improve the efficiency of financing schemes and to make them work in terms of actual household incomes as well as to stimulate savings directed to the housing sector. An important feature of the programme is the possibility of a greater diversity in housing finance services, away from the delivery of one type of housing product, the privately constructed house, towards funding for the enlargement and acquisition of used or rented housing.

One important development is the increasingly supportive attitude of the Mexican government to non-commercial bank system options that aim to build upon the experience developed by NGOs and civic organizations with progressive housing and alternative financial systems over the last decade (Comité Preparatorio de Mexico 1996). In the sections below we discuss four main non-commercial bank organizations. Two of these are saving and loan societies (Sociedades de Ahorro y Préstamo, SAPs) that are closely related to the formal financial system and limited mortgage brokerage companies (Sociedades Financieras de Objeto Limitado, SOFOLs). The other two organizations that support social initiatives are based upon self-financing and mutual assistance groups (Plan de Ayuda Mutua, PAM).

Saving and loan societies

Created in 1992, the Sociedades de Ahorro y Préstamo (SAPs) are an extension of the long-established *cajas de ahorro* and *cajas populares*, the informal saving systems used throughout Mexico (Galindo Guameros 1962). SAP members are united to improve their economic conditions based upon mutual trust to maintain deposits and to keep accounts. Legally registered as non-profit auxiliary-credit trusts, these so-called 'banks of the future' are really 'banks of the past'. In order to encourage their development the government has given various civil associations and co-operatives with experience in group savings projects two years in which to transform their operations to the new SAP scheme. To date, there are fifteen SAPs in operation, some of them newly created and others transformed from *cajas de ahorro*, with more than 1.5 million members and deposits worth approximately US$ 20 million.

The formation of SAPs presents savers with a range of advantages over conventional banking systems. SAPs are open to members of the formal and of the informal economy and to individual, family or group savers, which means that they have a potential market reaching about 80 per cent of Mexico's active labour force.[11] From initial evidence it would appear that savers feel comfortable with the SAPs due to the support of the Secretaría de Hacienda (Treasury) and controls imposed by the Banco de México and the Comisión Bancario. Administrative costs for SAPs, estimated at only 1.35 per cent for active capital at risk, are also attractive to potential savers and borrowers and there would seem to be a greater cultural appeal to working people who do not use the commercial banking system.

As a means to finance housing, SAPs can act in a number of ways. First, they can serve as a means to collect funds from savings groups and to provide loans under a self-financing scheme. Second, they can create co-operative societies and banks (see SOFOL below) with which members can open savings accounts in order to accumulate sufficient savings to qualify for finance. Third, the SAPs can use collective savings and co-ordinate with developers to construct members housing with control maintained by the local *caja popular* committee. An example of this last strategy is the Cooperativa de Consumo Caja Popular San Miguel de Mezquitán located in Tonalá, Guadalajara, which has built a housing development for its members. To qualify, a member must have a total monthly household income four times the level of total monthly payments, have been working for two years and make a small deposit. The *caja popular* will sell or transfer the mortgage to a FOVI account; thus replenishing its own revolving funds.

An extension of this approach is the new commercial banks being established in Ciudad Juárez and Tijuana with funds from the World Bank. The aim is for the banks to promote micro-enterprise efforts and form links with local groups and organizations that can provide land and construction services. In the case of Ciudad Juárez the bank is to be linked with a municipality-run trust known as Fideicomiso 16 that acquired 600 hectares of land to develop affordable housing. In this way the bank can provide credits for new construction, rehabilitation and mixed-use projects. The real difference here is that this is a local and not a national bank, and its focus is on micro-entrepreneurial efforts rather than national policy goals.

Limited-purpose finance companies

One of the most recent forms of non-commercial bank to be established in Mexico is the limited-purpose finance company, Sociedad Financiera de Objeto Limitado (SOFOL). The object of the SOFOL is to collect resources from public or private institutions by pooling financial instruments that are recognized as intermediate sources for financial transfers.[12] There is also the possibility of the further development of the SOFOLs if the government grants permission to accept direct deposit accounts from individuals, groups and households. In order to provide loans, a SOFOL needs a minimum start-up equity capital of US$ 1.6 billion that is equivalent to 25 per cent of the requirements imposed upon a commercial bank.

The advantage of the SOFOL lies in its flexibility and potential for adaptation to market and client needs, something the commercial banks do not appear to possess. SOFOLs can operate with minimum staff levels, little office space and existing market-information systems. Early experience reports low administrative fees amounting to 2 per cent of risk capital but high leverage with 1 peso of capital supporting 12.5 pesos of credit. A SOFOL can invest resources in high-yield instruments that have high liquidity in the financial markets. Administratively SOFOLs have demonstrated the feasibility of opening small collection offices in settlements which are staffed on collection days. This decentralized method of operation assures prompt loan payments, at periods as short as every fifteen days, again something the commercial banks find difficult to achieve.

One possible weakness of the SOFOL concept was the assumption that Mexico would have high rates of economic growth and that there would be a high demand for finance beyond the capacity of the commercial banks. Under this assumption, a lot of financial organizations made applications to the Secretaría de Hacienda for charters and by the end of 1994 the government had approved twenty-seven SOFOLs with another thirty in process. From the twenty-seven chartered SOFOLs, seventeen were specialists in mortgage processing, six were directed at small and intermediate enterprises, two at finance communication and transport projects, and two were created to fund rural development. However, due to the financial crisis of 1994–5 and the inability to get unsecured lines of credit, most of these institutions have experienced their own financial difficulties. At the end of 1995, therefore, changes were made to the permitted funding base to the SOFOLs after two of the largest societies, Financiero del Centro and Su Casita, obtained external support. Financiero del Centro signed a contract with the subsidiary of the Inter-American Development Bank for US$ 1 million of working capital and US$ 2 million over eight years, and Su Casita signed a contract with ICM Mortgage of Denver, a subsidiary of Pulte Corporation, as a foreign capital partner.[13]

At present, for the purpose of housing construction, a SOFOL is allowed to give loans for the purchase, construction or upgrading of housing. Given the correct circumstances a SOFOL could potentially finance projects previously funded by FONHAPO. Indeed, the SOFOLs already operate in similar ways to FONHAPO through flexible business practices and developing projects with clients, rather than waiting for the market to come to them. In Mexico City, for example, Su Casita has worked with a group of 400 households to secure construction loans for a housing project on land that the households bought collectively over a period of years. This type of service provides evidence to low-income households of the non-commercial bank sector's potential.

Self-financing and mutual assistance plans

Low-income households in Mexico have a long-standing scepticism of acquiring formal finance and considerable experience of various forms of self-finance in which people come together to save for the purchase of housing, cars or land. A study of land acquisition in irregular settlements found that the largest source of funds was personal savings, but that about two-thirds of households had recourse to other sources such as rotating savings schemes (*tandas*) or moneylenders, and about one-third sold furniture, received gifts or took extra jobs (Gareth Jones, personal communication). When finance was used, however, households were careful to use it wisely, either through acquisition of a plot in a single payment or in order to provide as large a deposit as possible in order to keep their exposure to a minimum. In the case of house construction, a separate study found that four-fifths of those interviewed had no recourse to formal finance even when they possessed full tenure rights, but again used personal savings (Varley 1987).

During the 1990s, the Mexican government attempted to formalize self-finance through requiring trusts to operate under the federal consumer laws. The trusts provide a system of reciprocal credit between savers who, individually, would be unable to make substantial purchases in the short term. Through a trust, a small regular savings contribution expands to a type of credit that is much larger than the savings amount. While the trusts operate with high charges, between 10–17 per cent of the cost of the purchase and 8 per cent of the administration costs, these are in lieu of interest and are far below payments under market rates. In order to retain the real value of the trust, the contributions are index-linked to a national bank rate (the CETES) or daily indexed-funds rate (the UDI). But, despite these safeguards, the trusts offer poor protection against inflation and members are vulnerable to non-payment by other participants.

To date, therefore, self-finance mechanisms offer limited opportunities to finance housing without additional seed capital from external sources. Accordingly the Mutual Assistance Plan (Plan de Ayuda Mutua, PAM) was devised as an assisted self-financing system for residents of irregular settlements to acquire land, housing and infrastructure rather than for daily consumption goods (Siembieda and López Moreno 1997). The PAM functions as a system of dedicated saving funds that consists of community-based groups of savers assisted by a collective agent such as a non-governmental organization (NGO). Government, private and donor sectors through savings-incentive contributions and seed-capital loans can also assist PAMs. Participants contribute a self-determined amount on a weekly or possibly daily basis over a savings cycle, and fund distribution is made by means of a lottery in which one participant receives the benefits, equivalent to the sum of the savings cycle, in the form of construction materials, labour or the down payment on land. No cash payments are made to lottery winners.

In many Mexican states occupation-based groups such as schoolteachers and electricians have organized to acquire land through special savings programmes. Frequently, they form civil associations and open bank accounts to which members contribute on a regular basis. In this fashion, members are able to acquire land at a fixed price and install simple infrastructure from the outset. In the state of Michoacán, parcels in a government-run territorial reserve programme have been purchased by groups who have then contracted a civil engineering company for land division and the design of community infrastructure. Such efforts are usually undertaken by households who earn more than twice the minimum salary and have the ability to participate in a multi-year saving scheme. For low-income households or those with irregular wages these schemes can work if savings requirements are set as realistic levels by the group.

Comparing the non-commercial bank alternatives

The non-commercial bank systems appear to be good ways to strengthen public savings because they possess high levels of social motivation with supportive organizational structures and offer affordable mechanisms to low-income groups. Their record indicates an ability to assist in organizing a flow of capital from a group in society that has traditionally been outside of the national financial system. Any of these non-commercial bank systems represents a natural match to the individual or group capacity for savings, and do not compete with commercial financial systems that concentrate and link capital to international markets.

The moderate fees and operating costs of these systems allow for short-term liquidity with the possibility of obtaining loans in the long term. Since they do not operate in the national and international capital markets, there are less stringent requirements to formulate and pay for a system of insurance and credit guarantees. Furthermore, the opening and control of accounts are done in a simple administrative way, the organizations work with a minimum of staff who do not provide a high level of skill and the management of joint contracts and fund collection is easy and fluid. These characteristics mean that members have a sense of participation in the system and belief that any benefits are returned to the members and not channelled to bank owners.

CONCLUSION

The challenge facing the Mexican housing finance system is how to integrate a variety of finance mechanisms (subsidies, incentives to save, flexible repayment schemes) at an administrative level (city or neighbourhood) that is appropriate to people's needs (land acquisition, upgrading, house purchase). Traditionally, the commercial banks and public housing finance agencies have been poorly suited to meet this challenge. They operate according to ever-changing national policy targets, delivering inefficient financial instruments that are neither cost-effective nor able to meet the needs of low-income households. Despite important reforms to the finance system that have created a secondary mortgage market and the resale of mortgage funds, many structural problems remain. Although Mexico possesses a sophisticated housing finance system, its track record at experimenting with progressive options and community-based finance models is poor.

By contrast, the recent experience of SOFOLs and SAPs appears to offer greater flexibility and more locally tailored financial instruments. The path of future reform should look to strengthen these mechanisms. One option is to consider ways to link the use of FOVI and INFONAVIT funds with the emerging non-commercial bank sector. Already, FOVI has established origination standards for credits initiated at SOFOLs and further effort to provide credits for resale housing is expected, more could be done (Zepeda 1996). It is possible that the state housing institutes might seek partnerships in the channelling of local resources to or through the non-commercial bank sector. Also, given some legal and operational encouragement from the federal government, the state housing institutes could create different types of financial products adjusted to regional economic conditions. These housing institutes could also bolster revolving finance schemes linked to collective saving organizations with matching funds. As shown by a number of projects in Jalisco, this kind of arrangement can allow an important change in the flow of resources and channel funds to achieve greater social coverage (Gobierno del Estado del Jalisco 1995).

The rise of the non-commercial bank sector and NGOs is a positive sign of a new capacity to adapt and be flexible and should encourage further reform and experimentation with the construction sector. Politicians, however, need to be made aware that the reform of the housing finance system can also provide non-housing benefits. Mexico's large, and possibly growing, national housing deficit marks one element of the polarization of society between those who can obtain decent housing and the rest who are forced to live in irregular housing. Such polarization will make political integration more difficult and lead to further social unrest in the cities. As a locally based solution that requires few external sources or subsidies, the non-commercial bank system can be a powerful tool in making the connection between national economic and political goals real and legitimate.

NOTES

1 FOVI was established under a framework of the Alliance for Progress with a US$ 20 million loan from the Inter-American Development Bank.

2 This was made possible by the increased leverage of the government on the banks following their nationalization in 1982 (Jones *et al.* 1993).

3 FONHAPO's 'success' has benefited from concession finance from the World Bank and substantial donations of land by state governments. More than one-half of the funds allocated to FONHAPO by BANOBRAS has come from grants or loans from the World Bank.

4 Although land and housing development in Mexico has most often been conducted through labour union branches, church groups and agrarian organizations, more radical social movements have also been important in Mexico City, Monterrey and northern cities such as Durango. Sometimes supported by NGOs or left-wing parties, but only rarely organized as collective or co-operative structures, these movements have been co-opted by the government through FONHAPO.

5 An example of such influence is the geographical allocation of INFONAVIT funds. The use of mandated payroll funds through a central agency was intended to allow poorer regions access to finance that would not be available from local collections. In reality, INFONAVIT housing finance has tended to be concentrated in industrial centres with strong union representation.

6 In the important state of Jalisco, for example, the Association for the Production of Housing (AJPROVI) found a significant divergence of costs between public financed and private developments in 1995 and 1996. Despite attempts to reduce costs through the construction of medium-density, mixed four-level apartments and single-family houses located in the larger cities, this divergence appears to be widespread.

7 There is some difficulty with interpreting data on FONHAPO as its portfolio includes small loans for housing improvement and, following the 1985 earthquake, an inheritance of 42,000 units from Renovación Habitacional Popular. Since 1988, FONHAPO has been downgraded by President Salinas's 'Solidarity' programme that claims to have built or improved 400,000 units.

8 A number of banks had adopted low-interest starter loans, such as Espacio introduced by Banamex, that were attractive to many buyers and provided tax benefits to the banks

9 Housing finance institutions were subject to deregulation as a condition of World Bank loans in 1989 and 1993, and in the course of the preparations for the North American Free Trade Agreement (NAFTA).

10 A part of this strategy is Programa Social de Ahorro para Vivienda (PROSAVI) whereby workers save for a down payment and then bid for loans based on the size of the down payment. FOVI can grant successful bidders additional subsidies and favourable interest rates.

11 Data provided by Victor Villa, former bank administrator and now advisor for non-bank systems, Guadalajara, June 1996.

12 Most SOFOLs are linked to FOVI programmes in order to develop projects eligible for additional finance and participate in SUBASTA, and to operate as a primary depository for government trust funds.

13 Pulte is one of the largest home-builders in the USA and one of a small group of foreign firms building housing in Mexico.

6

FORMAL HOUSING FINANCE AND THE ELDERLY IN SINGAPORE

Kwame Addae-Dapaah
National University of Singapore

INTRODUCTION

Housing in Singapore means more than the provision of shelter with all the requisite amenities. It also represents an important status symbol and investment. Indeed, it is commonly seen as a hallmark of achievement that is encapsulated in the 'Five Cs', namely Cash, Credit card, Condominium, Country club membership and Car. The fact that a large proportion of Singaporeans currently lives in high-standard accommodation can be largely attributed to government intervention. That this intervention has been so successful is remarkable given that at independence in 1959 the country had 'one of South-east Asia's largest slum and squatter populations' (Wong and Yeh 1985). Then a quarter of a million people lived in badly degenerated slums while a further one-third of a million lived in squatter settlements.

By 1990, 88 per cent of homes in Singapore were owner-occupied (Addae-Dapaah and Leong 1996a). This has largely been due to the efforts of the Housing Development Board (HDB), which is the major supplier of public housing in the country. Housing 86 per cent of Singapore's population, the HDB is funded predominantly from government housing development loans and mortgage finance loans (US\$ 5.9 billion and US\$ 12.4 billion respectively at the end of 1995). The HDB initially provided housing solely for low-income groups, a policy that was changed in the 1970s due to escalating property prices which priced many middle-income groups out of the private market. Consequently, HDB began to cater for middle-income groups as well.

While this widening of the HDB's target group has been impressive, it has not resolved the housing problems besetting all sections of the population. The elderly, in particular, have been marginalized by government housing schemes despite the fact that Singapore has the fifth-fastest ageing population in the world: between 1990–2010, this population will grow by 14–26 per cent (Kua 1994). Housing needs change over a person's life-course and elderly populations often find themselves facing uncertain housing conditions at a stage when they are least able to deal with them. One particular problem facing this section of the population is access to housing finance. For most elderly households, there are two main housing alternatives: namely, to 'stay put' or to purchase sheltered housing (Flesis 1985). Both of these options have significant financial implications for households either unable or unwilling to enter into the new finance arrangements required by policy frameworks which set out to maximize independence during old age (Mackintosh *et al.* 1990, Heumann and Boldy 1982, Newcomer *et al.* 1986, Katsura *et al.* 1989, Rose 1982).

An important dimension to the provision of housing and housing finance for the elderly is the changing demographic and gendered structure of this population. The elderly population in Singapore reveals a gender bias towards women. Among those aged 60 and over, there are 862 men for every 1,000 women and the gender divide becomes even more pronounced after the age of 75 when there are only 694 men per 1,000 women (Department of Statistics 1996). More than 37 per cent of women over the age of 75 are heads of households. Elderly women face specific housing problems. Across the developing world, these women have traditionally lived in extended households that may be disintegrating in response to modernization (Yen and Keigher 1992). Moreover, the fact that women, on average, earn less than men throughout their working lives means that they are ill-prepared for the financial situation they may face in old age. Women are less economically active than men in Singapore (50 per cent work as compared to over 78 per cent of men) and are also less well off with a median monthly income of US$ 921 as compared to US$ 1,315 for men. By the time the population is above the age of 70, only 23.5 per cent of women work (compared to 76.5 per cent of men) and 88.2 per cent of them earn less than the national median monthly income of US$ 1,140 (compared to 69.8 per cent of men).

In contrast to countries in the developed world which have ageing populations and where a variety of housing and finance schemes have been devised to meet the specific housing needs of the elderly, Singapore has only recently begun to deal with its problem. This chapter aims to assess the effectiveness of recent policies in resolving the housing finance crisis facing the elderly. It begins by critically reviewing housing policies which have appeared to provide Singapore with enviable housing conditions, but which have created limited housing options for the elderly and have led to a widening housing finance gap. It will then go on to evaluate more recent financial innovations such as the Home Equity Conversion Scheme and the Creative Housing Finance Scheme, and more specifically the Reverse Mortgage Scheme. The receptivity of the elderly to these schemes will be examined using data from a 1994 survey of 455 people (40 per cent of whom were women). A subsequent survey in 1995 (in which a further 180 people were interviewed) provides information on the acceptability and popularity of retirement villages among the elderly.

HOUSING FOR THE ELDERLY IN SINGAPORE

The housing options open to the elderly in Singapore are: to continue to live in their existing accommodation; to move to live with their children; to occupy government or private housing specially designed for the elderly; or to live in retirement villages. The key to realizing any of these alternatives is the availability and management of finance.

The 1990 census revealed that the least favoured option among the elderly is to live in institutions such as nursing homes and sheltered housing which provide accommodation to less than 1.5 per cent of the population (Lau 1992). This reluctance can be largely attributed to the social stigma attached to these institutions in a country where parents have traditionally been able to rely on their children for support in old age.[1] In general, 62 per cent of the respondents of the 1995 survey preferred to live in ordinary housing, 26 per cent in retirement villages, 9 per cent in sheltered housing and 3 per cent in institutional homes.

However, given the greying of the population of Singapore as well as changing socio-cultural values, there are tremendous opportunities for profit-making in the provision of 'affordable' retirement villages. This option is being actively promoted by organizations such as the Singapore National Co-operatives Federation and NTUC Income Insurance Co-operative Limited, both quasi-government institutions. Even though this research has revealed that the elderly prefer to live with their spouses and children, younger, wealthier and more educated respondents are generally more receptive to the idea of living in retirement villages. The trend is towards a retirement village lifestyle for the ambulant elderly who require special housing.

Despite the growing acceptance and availability of housing alternatives (see below), 65 per cent of the elderly prefer to live with or close to their children (HDB 1993). Although it is the preferred option among the elderly, only 36 per cent of young Singaporeans support the idea of living with their parents after marriage (Zaobao 1988). Increasingly, therefore, the most viable housing option for the majority of the elderly is to live 'alone' in their existing house (which may be unsuitable for old age and require substantial modification) or to rent/buy public or private housing which has already been modified for old-age needs.

Public housing in Singapore is predominantly produced by the HDB. The bulk of HDB housing has failed, until recently, to provide accommodation suitable for the specific needs of the elderly. Most HDB housing, for example, does not provide sufficient 'environmental leverage': the positional advantage and means to promote freedom of movement which is vitally important for the elderly to compensate for their reduced mobility. Moreover, much of this accommodation does not incorporate healthcare facilities nor does it take into account locational features which are crucial to the elderly as their life space contracts due to reduced mobility (Addae-Dapaah and Leong 1996a). In recognition of these shortcomings, the HDB and the Ministry of Community Development have recently begun to rehabilitate the flats of elderly tenants to meet their special requirements. Furthermore, the HDB has introduced 'special designed flats' which are amenable to elderly lifestyles in the newer housing estates. These schemes serve to enable the elderly to live with their children under one roof within the same unit – a strong preference among this population, as shown above.

While these changes have addressed the problems of design, they have left largely unresolved the question of affordability and the availability of housing finance required to either purchase or rent a specially modified house or to undertake modifications to an existing home. It is to these problems that we shall turn now.

Housing finance and the elderly

The Singaporean population has the choice of acquiring housing mortgages from either private or public finance institutions like the HDB. The majority of purchasers of HDB units acquire their housing finance from the HDB in order to capitalize on a sale price and mortgage interest rate that are both subsidized.[2] Furthermore, purchasers do not have to sacrifice their disposable income for the acquisition of HDB property by drawing upon finance from the 'approved housing scheme' run by the Central Provident Fund (CPF). The CPF operates through mandated contributions of 20 per cent of the gross salary of employees which are matched by a 20 per cent contribution from employers (Kim 1997). Under the CPF's 'approved housing scheme' members can utilize part or all of their contributions for the outright purchase of HDB flats or for the payment of monthly instalments. With the CPF capturing 40 per cent of gross salaries, the contribution of an average purchaser is more than enough to meet monthly repayments. As shown in Table 6.1, 84 per cent of households can afford to purchase high-quality four-room flats at the minimum price through their CPF contributions alone.

This innovative form of housing finance has significantly contributed to the high levels of home-ownership in the country and underpinned the buoyancy of the residential property market since 1985. Between 1985 and 1995, annual withdrawals from CPF savings for the purchase of residential properties rose (Figure 6.1). By 1995, 62.3 per cent of all employees who were aged 21 and above used their CPF savings to finance purchases of both public and private housing. For public housing alone, the figure was 90.7 per cent (CPF 1995).

However, these forms of housing finance have not provided a suitable solution to the housing problems of the elderly for a number of reasons. First, the lending criteria of both public and private institutions are heavily weighted against the elderly. Most financial institutions prefer to lend to borrowers below the age of 45 and require borrowers to have repaid their mortgages by the time they are 60. Second, the elderly fail to qualify for conventional mortgages as they do not have the means to service repayments without regular incomes. Even if the elderly do successfully acquire mortgage finance, their age prevents

Table 6.1 Percentage of households able to purchase HDB housing with CPF contributions

Type of flat		Price (US$)[a]	Mortgage (US$)[b]	Monthly amortization (US$)[c] 20 years	Monthly amortization as % of average household income at		% of population
					1990 (US$ 2,210)[d]	1995 (US$ 3,190)[e]	
4-room	Minimum	51,500	41,200	241	10.9	7.5	84.0
	Maximum	88,400	72,130	413	18.7	12.9	70.4
5-room	Minimum	89,250	71,400	417	18.9	13.1	70.4
	Maximum	140,500	112,500	657	29.7	20.6	49.8
Executive	Minimum	157,300	125,840	735	33.3	23.0	42.6
	Maximum	212,000	169,550	990	44.8	31.0	27.8

Notes:
a Minimum and maximum price figures are taken from HDB *Annual Report* 1994/95.
b 80% of price of flat. A lump sum down-payment of 20% of the price is a prerequisite. However, savings may not be used in the CPF to pay this lump sum.
c Mortgage is amortized at 3.58% per annum, which is the current interest rate that HDB charges on mortgage loans to Singaporeans.
d Based on household income figures from the 1990 Population Census.
e Based on estimated household income calculated by the author from available statistics.

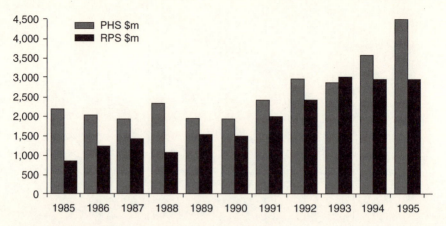

Figure 6.1 CPF annual withdrawals under PHS and RPS
Source: NHT's 1995 Mortgage Master File.

them from amortizing the mortgage over a sufficiently long period of time to keep their payments at an affordable level. Third, the capital value of property already owned by elderly households is insufficient to finance a move to appropriate accommodation. Even among home-owners who have paid off an earlier mortgage, only about 13 per cent possess a potential equity greater than the US$ 420,000 required to purchase a new private retirement home.

Fourth, the escalation in property prices (in both public and private sectors) has forced many middle-income households, let alone elderly households, out of the housing market.[3] It is estimated that private property prices increased two-fold between 1991 and 1995 and cumulatively private property prices increased about 72.5 per cent more than income between 1980 and 1994. Even HDB housing (especially the higher-standard units) has experienced sharp increases in sale price. Between 1994 and 1995, 29,280 HDB flats changed hands on the secondhand market with four- and five-room, and executive flats selling for average prices of US$ 210,000, US$ 310,000 and US$ 405,000 respectively in 1996.

These rising prices have meant that property cannot be financed solely through CPF contributions. It is estimated that the average Singaporean household purchasing a five-room or an executive flat on the secondhand market may have to contribute an additional US$ 210 or US$ 650 per month respectively. This works out to be 26.7 and 41 per cent respectively of the gross average household monthly income. With an average price of US$ 523,000 for condominiums, affordability is stretched even for some of the top 28 per cent of households who earn more than US$ 3,500 per month. The higher-standard accommodation such as three- to five-roomed flats and executive condominiums which sell at a maximum price of US$ 210,000 are both out of the reach of elderly households through the CPF (Table 6.2).

Table 6.2 Schedule of monthly income from CPF balance at retirement

CPF Balance at age 55 (US$)	Monthly income (US$)[a]	Affordable housing cost (US$)[b]
34,892	201	64
69,785	387	129
139,567	774	258
279,135	1,548	516
418,702	2,322	774
558,269	3,096	1,032
697,837	3,870	1,290
1,046,755	5,805	1,935

Notes:

The CPF balance is shown here for US$ equivalents of 50,000, 100,000, 200,000, 400,000, 600,000, 800,000, 1,000,000 and 1,500,000 Singapore dollar amounts.

a. Assuming average life expectancy of 74 years, monthly income is computed for a remaining life span of 20 years from retirement at age 55 at savings interest rate of 3% per annum, monthly compounding monthly income = CPF balance $\times MC_{20 \times 12.3\%}$.

b Housing cost is assumed to be one-third of monthly income, a rule of thumb applied in Singapore.

Indeed, affordability is even a problem in the rental sector. In general, a household with a maximum income of US$ 5,580 can afford to rent decent housing subject to eligibility criteria. However, the elderly, who generally have lower incomes and whose CPF balance is below US$ 105,000 have to compete with younger applicants for the rental of two-room HDB flats at rents which consume a higher proportion of their monthly budget (Table 6.2). The option of renting a two-bedroom flat in a private retirement housing scheme in Singapore is estimated to cost at least US$ 1,520 per month, about ten times the sums that an elderly household might be able to devote to rent, unless their CPF balance is greater than US$ 768,000

Finally, in the absence of retirement homes at an affordable price, 87 per cent of the elderly have to stay put, a solution which has its own particular problems as many properties in Singapore are leasehold. HDB flats, for example, have a ninety-nine year lease and at retirement age most leases will have less than sixty years to run. Since current CPF regulations forbid the use of savings to purchase houses with less than sixty years of unexpired lease, most of the houses owned by the elderly have a reduced economic life of only a few years. This severely prejudices the resale of the houses of the elderly as no rational house-purchaser will buy a house which they may not be able to sell after a few years.

The lack of finance as well as escalating property prices and rents have resulted in a housing finance crisis for the elderly. This in turn has prompted the government as well as private-sector institutions to experiment with alternative finance schemes to enable the elderly to either move to new accommodation or to undertake modifications of their existing homes.

HOUSING FINANCE INNOVATION AND THE ELDERLY

With a high level of home-ownership, many households in Singapore have a lot of equity tied up in their houses. For this reason, home equity conversion schemes (HECSs) have been seen as an attractive option for providing the 'house-rich, cash-poor' elderly with an additional financial resource to meet their housing needs. A study by Leather *et al.* (1988) emphasizes the importance of using home equity for housing repair, maintenance, improvement or adaptation in old age. HECSs can be divided into three categories: mortgage annuity schemes, home reversion schemes and deferred payment loans. All three are designed to make home equity values and not household income the basis of eligibility for loans (Figure 6.2). Thus, they provide the elderly with an income from the equity tied up in the house while allowing them to continue occupancy.

By contrast to HECs, creative housing finance schemes (CHFSs) attempt to raise the affordability level of new homes for the elderly by lowering the price and/or the size of the initial financial outlay (Figure 6.3).[4] This can be achieved in a number of ways. First, land costs can be reduced if subsidized through land grants. Evidence from the United Kingdom reveals that the issuing of special land grants by local authorities can lead to a significant reduction in rent. This, in turn, can result in the provision of sheltered housing developments for the elderly at competitive prices (Mackintosh *et al.* 1990). Second, financial costs can be lowered by the adoption of unconventional financial strategies such as raising finance from future residents. This finance, often seen

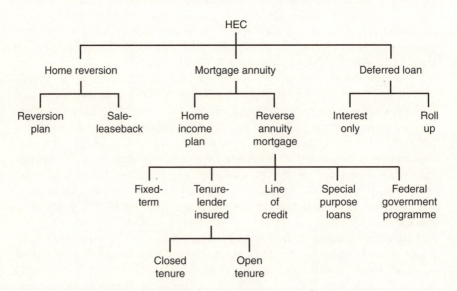

Figure 6.2 Home equity conversion schemes (HECSs)

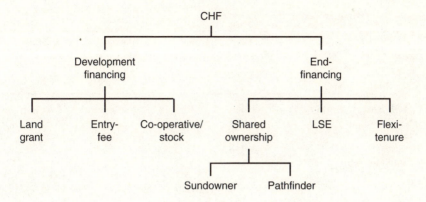

Figure 6.3 Creative housing finance schemes (CHFSs)

as an entry fee to the scheme, can be used to build an endowment fund for the project. Permanent funding can be achieved by new residents paying entrance fees for vacated housing while departing residents' fees are (partially or fully) refunded. Third, instead of a fee system, funding can be obtained from future residents through the formation of co-operatives by the sale of shares to residents. Sometimes referred to as co-operative and loan stock schemes, these schemes can be used to purchase sheltered housing through the provision of interest-free loans. Finally, shared ownership schemes, whereby residents part-own and part-rent a property, can make housing more affordable to the elderly. However, these schemes only work if the cost of renting is below the level of mortgage repayments and therefore usually require some element of subsidization.

In Singapore, a HECSs, or more specifically a 'reverse mortgage scheme', has been adopted by the NTUC Income Insurance Co-operative Limited which has identified a potential market of 9,000 elderly private home-owners and 20,000 elderly owners of three-room HDB flats. This is probably a conservative estimate as Table 6.3 shows that the potential market, especially from HDB flat-owners, could be a great deal larger. Reverse mortgage schemes have been offered by the NTUC to people over the age of 60. According to the company, a 60-year old person with a property worth US$ 350,000 who acquires a reverse mortgage can expect to draw a monthly advance of US$ 840 for the next twenty-four years. While this sum could go a long way towards resolving the housing finance problem of elderly households, the figure appears to be highly unrealistic. Given an equity of US$ 350,000, a loan/value ratio of 80 per cent and an interest rate of 6.4 per cent per annum (monthly compounding over twenty-four years), the monthly annuity is likely to be closer to US$ 410. Thus, the only way by which NTUC Income can pay an annuity of US$ 840 per month is to charge an interest rate of about 12 per cent per annum, monthly compounding.

Table 6.3 Potential home equity of the elderly

Type of property	Value (US$)[a]	Number of homeowners	% of homeowners[b]	Monthly loan advance (US$)[c]	Monthly loan advance	
					5.5%	6.0%
HDB 3-room	112,352	41,844	40.5	152	152	142
HDB 4-room	209,350	30,947	30.0	280	280	261
HDB 5-room/						
executive	357,292	13,438	13.0	479	479	446
Condominium						
(L99)	523,378	3,339	3.2	703	703	653
Terrace	976,971	7,171	6.9	1,311	1,311	1,219
Semi-detached	1,744,592	4,187	4.1	2,341	2,341	2,177
Bungalow	3,210,050	2,334	2.3	4,308	4,308	4,005
Total		103,260	100.0			

Source: Based on Figures in Department of Statistics (1996).
Notes:
a Average market value as of March 1996.
b Percentage of elderly home-owners over elderly heads of households.
c The advance is calculated based on the assumption that the mortgage has a fixed interest which is monthly compounding for a 24-year loan term at loan/value ratio of 80%.

The technical and practical difficulties of reverse mortgage schemes notwithstanding, as many as 44 per cent of elderly owners of private housing had not heard of the scheme in question. Eighty-one per cent did not want to have anything to do with such a scheme, while 19 per cent responded that they would consider taking a reverse mortgage if the terms and conditions were favourable. These findings suggest that the probable market size for reverse mortgages among elderly private home-owners would be about 2,567 retiree home-owners, most of whom would require to be convinced of the scheme benefits. The poor patronage of the scheme from this sector of the community is attributable to their fear of losing ownership of their flat *vis-à-vis* a bequest motive, expectation that children will take care of them and the complexity of the schemes. Typical responses from this group were that reverse mortgages were 'not necessary' and 'my children will take care of me'. It must be conceded that the probability of losing one's property due to default is real, especially when one is dealing with a private financial institution. Loss of ownership can be avoided, however, if one takes a reverse mortgage that guarantees ownership of property as long as one lives in it and as NTUC is a quasi-government institution, the probability of loss of ownership is reduced. This is reinforced by the fact that the Singapore government has shown itself to be committed to enhancing the assets of Singaporeans and therefore is unlikely to condone loss of properties to NTUC Income through reverse mortgages.

Unlike the elderly living in private housing, owners of HDB flats were more receptive to the reverse mortgage scheme. Although fewer than one-quarter of the respondents knew of the scheme, 23.7 per cent welcomed it (70 per cent of whom knew of the scheme). In part the greater popularity of this scheme among the elderly living in HDB flats is due to the fact that they are relatively less well off than those living in private accommodation. Thus, for example, only 15 per cent of this group felt that they would be able to cope with the cost of living in retirement and 85 per cent of subscribers to reverse mortgages would need extra finance to make ends meet. Assuming a loan/value ratio of 80 per cent and an interest rate of 5.5 per cent per annum (monthly compound) an elderly owner of a four-room HDB flat with an average value of US$ 210,000 could derive a monthly income of US$ 335 over twenty-four years through a reverse mortgage. Thus, a reverse mortgage could be a useful mechanism for assisting these elderly households. According to the survey, 34.8, 27.8 and 30.7 per cent respecively of owners of three-, four- and five-room and executive flats would take a reverse mortgage. These figures translate into a probable market size for reverse mortgages of 27,306 elderly HDB flat owners.

The main reason for the 23.7 per cent of HDB owners welcoming the reverse mortgage is the desire to be financially independent and not to be a burden to their children. Thus the quest for financial independence during retirement may induce more of the elderly to subscribe to reverse mortgages if the fear of losing ownership of their house or flat could be allayed. However, at the same time, this receptivity has to be weighed against a general wish among the elderly population to bequeath their equity intact to their children rather than pawning it for their own benefit (Addae-Dapaah and Leong 1996a). Furthermore, there are difficulties in advancing these schemes to owners who value them most. At present, HDB flat owners cannot take advantage of the reverse mortgage scheme because insurance companies do not accept their properties as investment assets and HDB regulations forbid the assignment of HDB flats to financial institutions.

CONCLUSION

Although the Singapore housing market appears to have been effective in resolving the housing crisis that plagued the country in the early 1960s, it has marginalized an important section of the population, the elderly. In recognition of this, both the government and the private sector have attempted to introduce diverse housing finance schemes that aim to assist the elderly to optimize their living conditions. Even though innovative housing finance schemes are fraught with legal, technical and social problems, they have the potential to alleviate the financial predicament of elderly Singaporeans.

To date, however, the increased incomes that these schemes offer have not resolved the housing difficulties faced by the elderly. This is due to the fact that income from schemes such as the reverse mortgage is periodic whereas elderly Singaporeans appear to want a lump sum of capital to adapt their existing homes. Until such schemes become established and adapted to local needs, it would appear that other measures will have to be taken in the short term. First, there are sale-leaseback arrangements whereby the HDB purchases the homes of the elderly at the market price, renovates them to suit the special needs of the elderly and leases them back at an appropriate rent. Second, the HDB may buy the flats of the elderly at the market price to provide them with the means to finance retirement village housing. Third, the government might reform existing regulations so that the saleability attendant to the unexpired term of the lease may be resolved by giving a life-for-ninety-nine-year lease to all Singaporeans who buy flats from the HDB for the first time.

Of course, whichever measures are taken to resolve the housing finance problem in Singapore, they do no more than mask a problem of saleability of the existing stock. High-rise blocks in Singapore have a life-span of fifty years and, assuming that the average Singaporean purchases a HDB flat at age 30 and takes out a reverse mortgage at the age of 60 for twenty-four years, the flat would be fifty-four years old by the end of the mortgage term. Thus, there is a very real danger that the physical obsolescence of buildings would render the property unmarketable and undermine the success of any innovative approach. Getting housing finance right for the potentially large reverse mortgage market offered by HDB flat-owners will require more than just financial innovation if the private sector is to be presented with viable opportunities to help the elderly.

NOTES

1 Indeed, 'abandoned' parents can seek legal redress via the 1995 Maintenance of Parents Act whereby parents can compel their financially able children to provide them with basic shelter.
2 HDB mortgages are extended for a five- to twenty-five-year period at a variable mortgage interest rate of 3.58 per cent per annum as compared to an average 6.5 per cent per annum from other financial institutions.
3 Attempts have been made by the government to tackle the issues of affordability, particularly in the private sector. The introduction of a capital gains tax on properties which are sold within three years of purchase to curb speculation and reduce prices, and 'caps' imposed by the Monetary Authority of Singapore (MAS) on the loan/value-ratio to 80 per cent. Furthermore, the government has made more land available for private housing development through the Urban Redevelopment Authority (URA 1996).
4 There are various types of CHFS including co-operative and loan stock schemes, end-financing schemes, shared ownership, sundowner schemes and pathfinder schemes.

7

BETTER A 'HUT' ON THE GROUND THAN A CASTLE IN THE AIR

Formal and informal housing finance for the urban poor in India[1]

Robert-Jan Baken and Peer Smets
Institute for Housing Studies, Rotterdam and Vrije Universiteit,
Amsterdam, The Netherlands

INTRODUCTION

This chapter deals with the formal and informal housing finance markets in two Indian cities, Visakhapatnam and Hyderabad, in the state of Andhra Pradesh. The chapter concentrates on how formal and informal housing finance relate to the housing conditions of the 'economically weaker section' (EWS) households, defined as having an income of less than Rs 1,250 per month (US$ 41), which is regarded as barely sufficient to cover basic needs.[2] This group represented approximately 26.1 per cent of the urban population in 1987–8 (Dandekar *et al.* 1993: 19). Our aim is to examine to what extent each finance market is compatible with the survival strategies of the low-income households by introducing the concept of the 'finance gap' and the implications for household debt and displacement. We argue that the form of house construction and the financial arrangements which are part of public housing programmes do not match household survival strategies and that the terms and conditions of informal housing finance are generally more adequate.

The material for this chapter was gathered through two independent research projects conducted in 1993–4. The first was on land and housing market development in Vijayawada and Visakhapatnam. Material from Visakhapatnam, one of India's fastest-growing cities with a growth rate of more than 60 per cent per decade and a 1991 population of 1.05 million, is used in this chapter to illustrate formal housing finance. The project included a survey of sixty-six settlements, selected on the basis of location, method of land delivery, age, and level of consolidation, as well as the form of government intervention (sixteen settlements had permanent houses built through an urban housing pro-gramme). While conducting the surveys, in-depth interviews were held to discern the impacts of the housing programme on the financial situation of households and the peculiarities of implementation in particular areas. A selec-tion of these interviews are presented as case studies in this chapter. The second project focused on housing finance in Hyderabad and Secunderabad. Material from Hyderabad, a city with a growth rate of 57 per cent from 1981 to 1991 and a metropolitan population of 4.3 million in 1991, is used to illustrate informal housing finance. Data were gathered from a survey of thirteen slums where no housing programmes have been initiated. Supplementary information was obtained from talks with local leaders and financial agents.

THE EVOLUTION OF FORMAL HOUSING FINANCE IN INDIA

In the decades following independence, the housing strategy of the Indian gov-ernment was to focus on housing low-income people in 'decent standard housing', both through conventional housing projects and slum clearance-cum-resettlement programmes. In 1970, the Housing and Urban Development Corporation (HUDCO) was created in order to undertake public housing and urban development programmes and to finance the activities of state governments operating in this field. By the late 1970s, the policy focus had shifted toward a dual approach of slum improvement and the large-scale delivery of rudimentary serviced land, housing finance and other housing inputs so as to arrest further slum formation. By 1988, however, the policy environment changed once more, towards the promotion of a liberalized hous-ing finance market. Of major importance was the establishment of the National Housing Bank (NHB), a subsidiary of the Reserve Bank of India, to support private housing finance corporations and encourage them to move their lending operations down-market by distributing small loans and adopt-ing a sliding scale of interest rates (Smets 1997).[3] The idea was that the construction of housing would be left to the low-income groups with financial assistance from government and private institutions. By 1994, housing finance represented the largest share of the public housing budget.

Despite some degree of liberalization, eligibility for a housing loan through a finance corporation remains restricted to salaried employees in the formal public or private sector, and to those self-employed people able to provide tax records. In this respect it should be noted that housing finance delivery is dis-criminatory, particularly against women, in a number of ways. Women are under-represented in the formal sector, which is mainly the preserve of men, with the result that women are excluded from private housing loans. Moreover, in practice, women can only gain access to housing finance through a husband who has to provide proof of income. This has the effect of excluding most female-headed households. As a consequence most women and almost anyone unable to prove formal employment status have to rely on the informal housing finance market where 'all financial transactions, loans and deposits, occur outside the regulation of a central monetary and financial market authority' (Adams and Fitchett 1992: 2, Srinivas and Higuchi 1996). To illustrate how formal housing finance is incompatible with the financial requirements and housing needs of low-income households, below we look in detail at the experience of the Weaker Section Housing Programme in Visakhapatnam.

The Weaker Section Housing Programme in Visakhapatnam

In this section, we focus on the impact of the Weaker Section Housing Programme (WSHP) which, since its initiation in 1979 has accounted for approximately 50 per cent of the total housing and urban development budget of the government of Andhra Pradesh. In 1979, 170 slums in Visakhapatnam were included in an Urban Community Development (UCD) project with local government finance and professional assistance from the Urban Community Development Project in Hyderabad. In the initial phases of the WSHP some 400–500 loans were arranged through private banks with the remainder sanctioned by the Andhra Pradesh State Housing Corporation (APSHC), and later by an offshoot, the Andhra Pradesh Urban Development and Housing Corporation (APUD&HC). In turn, these organizations received finance for loan distribution from HUDCO, which also deputized a number of executive staff to the municipal corporation which acted as the implementing agency.

In 1988, the United Kingdom's Overseas Development Administration (ODA) began to extend finance to the programme, with the consequence that a hitherto modest project focusing on community development and welfare services became an integrated slum improvement project covering 198 slums and including infrastructure, health and economic support. Although not a central part of the ODA project, the distribution of housing loans formed an important component of the slum improvement programme. Between 1979 and 1990, 12,361 loans were distributed to slum-dwellers in approximately seventy-five slums and, in 1994, proposals concerning a further 2,391 loans in twenty settlements were at various stages of approval. By 1994, ODA had spent some Rs 270 million on the project.

The implementation of the programme has been a long and complicated process. First, the District Collector and the Municipal Commissioner have to select the settlements to participate under the technical guidance of the UCD project staff and the manoeuvring of local leaders who frequently regard the housing loans as a political commodity. Second, after the identification of settlements, the APSHC/APUD&HC have to submit project proposals to HUDCO in Hyderabad for technical and financial approval. In doing so the project has to adhere to some basic HUDCO guidelines, namely: a minimum plinth area per unit of 13 square metres and a housing design with a permanent roof; a loan ceiling of Rs 19,500; a maximum cost per unit of Rs 26,400; a maximum repayment period of fifteen years for the Corporation and twenty years for the households; and an interest rate of 9 per cent per annum for the Housing Corporation. Within these guidelines, the Housing Corporation is allowed to devise its own cost estimates taking into account the limits on plot size. In 1994, the State government fixed the loan level at Rs 11,700, plus a Rs 1,000 grant and an obligatory contribution of Rs 300 from the beneficiary household, making a total potential loan of Rs 13,000. This was supposed to be sufficient to enable households to construct a simple permanent house of uniform design with a kitchen, living room and a small veranda.

Once the loans have been sanctioned by HUDCO, disbursement takes place through the phased distribution of cash and building materials. Repayment charges are calculated at a flat interest rate of 10 per cent over twenty years with 240 monthly instalments of Rs 98. At a settlement level, the UCD project officials make a list of beneficiaries and the local branch of the State Revenue Department issues conditional land titles (*pattas*) which are used as collateral for the loan. Finally, the assistant engineers and the work inspectors enter the settlement to prepare the phased implementation of the loan programme and guide the beneficiaries in the construction process.

The impact of the Weaker Section Housing Programme can be illustrated with two household case studies in Siva Nagar, a slum in Visakhapatnam which, after a fire in May 1988, was included in the programme.

Suresh and Radha

Suresh, Radha and their son live in a small hut near Siva Nagar on what is said to be government land. Suresh is a self-employed carpenter, working on a contract basis. If he doesn't fall ill or face any other unforeseen problems, he can earn about Rs 1,000 per month. In 1988, Suresh got a WSHP loan and with it the chance to build a 'real' house, which seemed to be a dream come true. In accordance with settlement reblocking, Suresh was required by the Urban Development and Housing Corporation to demolish the thatched house in which he and his family lived and construct a temporary hut while building work began.

Soon after Suresh began building work, however, it became clear that it would be impossible to complete house construction according to the prescribed plan without tapping additional sources of finance. A neighbour, Subba Rao, who lives in the same lane, lent Suresh some money at an interest rate of 4 per cent per month. Moreover, in order to acquire the cement bags and cash to which Suresh was entitled after the completion of each construction phase, bribes had to be paid to the assistant engineers supervising the construction work. Furthermore, income had to be foregone as a result of the days spent in government offices trying to get cheques signed by the authorities. With the

---

additional costs of the temporary hut and the administrative charges, it is calculated that Suresh spent some 21 per cent of his loan on non-construction
expenses.

By 1990, when Suresh had managed to construct the house up to roof level,
he had borrowed Rs 10,000 from the neighbour. This meant that Suresh and his
family had to spend 40 per cent of their monthly income on interest for the
informal loans. On top of this, when Suresh and Radha's son fell ill and had to
be treated in the government hospital, a considerable amount of money had to
be found in order to ensure reasonable treatment. The family's financial situation
worsened and finally they went to a local broker to sell their incomplete house.
Suresh and Radha continue to live in the temporary hut.

Vasu and Padma
Vasu, his wife Padma and their children rent a unit in a thatched house bor-
dering Siva Nagar. Vasu is a rickshaw puller and his wife works as a housemaid
which, collectively, provides an average monthly household income of about Rs
1,100. Vasu and Padma have gone through a similar process to Suresh and
Radha, with the difference that they have been able to complete their house. On
top of the government loan, Vasu has borrowed Rs 20,000 from a moneylender
at a monthly interest of Rs 600. When the household budget became stretched
it was decided to rent out the house and find a more modest thatched unit for
themselves. In this way the household saves Rs 250 per month, although it
seems unlikely that Vasu and Padma will ever be able to afford to move back to
their WSHP house.

Therefore, although households in the programme settlements are better off
than those in squatter settlements generally, these cases show that, contrary to
programme guidelines, many are unable to bridge the significant finance gap
between the real costs of construction and the housing loan.[4] This results in
either the displacement of the poorest households such as Suresh and Radha
who have had to sell their house or Vasu and Padma who have had to sub-let.

The finance gap

According to the prevailing market prices in Visakhapatnam in 1994, the
costs of constructing the minimum model unit required by the APSHC was
Rs 28,250. By contrast, the official unit costs covered by the loan are
Rs 13,000, a sum which the authorities know is unrealistic and leaves the
supposed beneficiaries with a substantial finance gap. Table 7.1 indicates
that for Suresh and his family only 42.5 per cent of the programme expenses
were covered by the loan. Since, in practice, many beneficiaries do not get the
full amount they are entitled to and since this amount includes a financial
contribution of their own, the finance gap presented in Table 7.1 is a conservative estimate.

Table 7.1 The finance gap, 1993–4

Expenses (in Rs)		Sources of finance (in Rs)[a]	
Admission fee	61	Loan	11,700
Administrative charges	260	Grant	1,000
Construction programme house	28,250	Beneficiary contribution	300
Construction temporary hut	1,500		
4–6 working days lost	200		
Bribes	300		
Photocopies/photographs	25	Finance gap	17,596
Total	30,596	Total	30,596

Note:
a Finance sources formally provided by the housing programme.

Moreover, there is evidence that the finance gap has widened over time. With the assistance of a civil engineer we calculated the average cost of constructing a programme house over time by deflating 1993–4 market prices using the official Standard Schedule of Rates for construction inputs weighted by the conventional deviation between market and government-estimated prices. The calculation gives a good indication of construction costs in different years and reveals that, whereas in 1979 the loan would have covered approximately 44 per cent of costs, by 1983 the figure was 55 per cent and thereafter the proportion of costs covered by the loans steadily declined to 42.5 per cent by 1994.

Impacts of the housing finance programme: displacement and debts

With such a large finance gap between construction costs and housing finance provision a survey of 2,335 households in thirty-four programme settlements sought to discover whether the introduction of housing finance had provoked increased levels of displacement and household debt. The results reveal that approximately 70 per cent of the house units are owned by occupant-beneficiaries and about 15 per cent by occupant-buyers. The remaining 15 per cent are owned by absentee landlords: 8 per cent buyers and 7 per cent beneficiaries. Looking specifically at the households living in the earliest programme settlements (1983–4), one finds that the share of the beneficiaries continuing to live in the original house is slightly less than 60 per cent.

To support the suggestion that housing finance has caused significant levels of displacement, the residential status of households in programme settlements was compared to that in settlements in which the finance programme was not applied. Our analysis indicates that population turnover is higher in the programme settlements. In twenty-eight non-programme squatter settlements the rate of housing turnover over an average period of thirty years was 5.7 per cent, far below the programme figures. Furthermore, data from fifteen housing programme settlements were used to compare pre- and post-programme turnover. The analysis found that after the introduction of housing finance the yearly turnover increased by 600 per cent, which can only be attributed to the impact of the programme. Indeed, according to our survey, a considerable number of households sold out even before starting construction. Yet, the sale of programme units did not imply substantial financial gains for all households and usually amounted to just a few thousand rupees.

As already pointed out, the construction of programme houses requires substantial additional financial contributions from the beneficiaries. Given the fact that a large share of the beneficiary households have to subsist on low and insecure incomes, common sense suggests that financial problems are the main cause of the widespread selling and renting of programme units. Indeed, for the poorer households who do not own significant assets such as gold or land in their family villages, but who still want to complete their programme unit, the only alternative is to obtain a sizeable informal loan, usually from a moneylender. For 63 per cent of the owner-occupant beneficiaries who managed to hold on to their houses there is evidence of considerable construction debts (median Rs 15,000). About 60 per cent of households mention moneylenders as the main source of additional housing finance, with typical interest rates above those in the formal private sector: 14.5 per cent of borrowers pay 3 per cent or less per month, while 35 per cent pay 5 per cent or more. Consequently, low-income households in the programme were found to be paying on average 30–40 per cent of their monthly income on interest. For them, the programme functions like a debt trap; provision of housing finance has forced households to take on extra levels of debt.

A comparison of total household debt between programme settlements and twenty-four invasion settlements shows that only 16 per cent of households in the former had no debts related to construction compared to 26 per cent in the invasion settlements. The invasion settlement households also held lower absolute debts: 43 per cent had a debt of Rs 5,000 or less, with a median debt of Rs 6,000, compared to only 10 per cent of households in programme settlements owing less than Rs 5,000 and a median debt of Rs 15,000, a sum larger than the original loan! In about 20 per cent of cases, households sought to cover the finance gap by renting their houses usually under a no rent-no interest agreement: a mutual agreement between the tenant and the landlord in which tenants provide interest-free (housing) loans to the landlord and in return do not have to pay rent. Across the board, the mounting debt burden among original beneficiaries still living in the programme houses has had the effect of lowering the disposable household income available for basic needs with obvious consequences for malnutrition and health standards (Asthana 1994).

Given such findings one would expect displacement to affect mostly poorer households. In-depth interviews with political and community leaders in eight programme settlements confirmed that the displaced belonged to the poorer sections of the slum population and that the pattern described in the case studies was fairly typical. One notable group among the displaced population is low-income female-headed households, who represent 14 per cent of displaced people compared to 7.7 per cent of programme households and 7.9 per cent of households in all squatter settlements. The vulnerability of female-headed households to displacement was heightened by their lower incomes compared to male-headed households: 45.4 per cent of female-headed households earned a monthly income of less than Rs 1,000 and 10.6 per cent earned more than Rs 2,500, compared to 24.8 per cent and 23.5 per cent among male-headed households.[5]

Although female-headed households are not found disproportionately in any particular settlement type, they do possess worse housing conditions than male-headed households wherever they are located. As housing design in programme settlements is prescribed by the UCP regardless of household structure, our survey looked at the quality of housing in squatter settlements. The findings reveal that whereas 27.8 per cent of female-headed households lived in makeshift huts, only 14.7 per cent of the male-headed households lived in this type of housing. Moreover, the share of households living in units with permanent walls made of brick or cement block was considerably higher among male-headed households (26.1 per cent) than among female-headed (14.4 per cent). This suggests that whether female-headed households stay or leave the programme settlements they are likely to live in the worst housing conditions.

Cases of beneficiaries selling houses in programme settlements in order to move up-market are rare and counter-intuitive. In terms of affordability, the nearest up-market alternative to a programme house would be an unserviced plot in an illegal subdivision which, in Visakhapatnam, is a segment of the land market that almost exclusively caters for middle- and higher-income groups. The price of an 80 metre square plot on a reasonable non-prime location would be equal to or more than the price of an average programme house. If the household does not have a substantial debt burden, buying such a plot might be possible so long as additional funds to finance construction can be found. It seems unlikely, however, that displaced people from programme settlements could fund such a move as most are already financially dependent upon informal finance. The survey results show that whereas most programme households depend on moneylenders as the main source of housing finance, the arrivals depend upon savings and assets. This suggests that, although income differences may not be that marked, the financial position of buyers is better than that of the original beneficiaries.

Community impacts

So far, we have mainly focused on the impacts of the WSHP at a household level. Compared with the informal finance mechanisms which we describe later, the wider community implications of the WSHP call for explicit attention as the housing programmes in Visakhapatnam involve whole communities and imply far-reaching changes to the physical and social lay-out of slums. As research into the impacts of the UCD project in five slums in 1988 showed, one consequence of the programme was a general increase in the political manipulation of slum-dwellers (Asthana 1994). The power and prestige associated with securing housing improvements acted as an invitation to political leaders who saw opportunities to build patronage networks by controlling the distribution of loans and plots. Indeed, the process of enumeration and the selection of beneficiaries is generally undertaken in an environment of mistrust with everybody suspecting others of manipulating the lists. During fieldwork we observed community organizers working for the slum improvement project immersed in numerous disputes between the beneficiaries and political leaders.

The introduction of housing finance programmes into settlements has the potential to realign political structures, often to the disadvantage of women. Asthana (1994) found that a strong community organization, largely based on local women, was disrupted by the implementation of a housing programme. Male Congress Party leaders took over key positions, shifting priorities from 'soft' community concerns such as welfare and healthcare issues to 'hard' issues such as housing. On top of this, the area witnessed widespread displacement with the leaders often taking on the role of brokering house sales and earning about Rs 1,500 for each sale (Asthana 1994: 65). In 1988, about nine years after the start of the housing programme, only 48 per cent of the programme houses were inhabited by the original beneficiaries. Continuity in community affairs was lost and the most important part of the target group, the EWS, had moved out. As one respondent explained:

> Since constructing the *pakka* houses everybody has got too big for their boots and the togetherness we felt has been lost. We used to raise one Rupee every month from every household to pay for the schoolteacher. Now we can't even get that. We have lost the earlier generation who were poor but honest. Now we're just a community of strange faces.
>
> (Asthana 1994: 68)

In addition to problems of debt, displacement and social cohesion, the way the housing programme operates has implications for the sustainability of the WSHP. The WSHP loans are not perceived in the same way as an ordinary loan, but as a gift from local politicians who, especially at the higher levels of party and faction hierarchies, incessantly play down the need for adequate recovery and, indeed, force 'their' officers not to take action in this respect. Low-income households, therefore, speculate on whether the populist moves of important politicians will free them from their obligation to repay loans.

Within such a politicized environment loan recovery is poor. Over the period 1983–4 to 1992–3 the loan recovery performance of the WSHP in urban and rural Andhra Pradesh was 5.1 per cent (Government of Andhra Pradesh 1994: 55).

Thus, the WSHP presents a number of paradoxes for urban development in India. First, it would appear that the target beneficiaries of the housing finance programme are the most likely to be displaced whereas those who manage to hold on to their houses should not have been included in any case as their incomes exceed the prescribed limit. Second, individual households are increasingly placed in debt while the programme itself achieves a cost recovery rate that makes the project financially unsustainable without injections of new money or subsidies.

INFORMAL HOUSING FINANCE FOR LOW-INCOME PEOPLE IN HYDERABAD[6]

As the foregoing discussion makes clear, low-income people have no other option than to rely on informal finance and live in so-called slum settlements. In Hyderabad, in 1993–4, a total of 662 such slums housing 205,200 people were officially recognized within the corporation limits. A survey carried out in a representative selection of thirteen slums found that 61 per cent of the housing stock was built in stages by owner-occupants. Most new house-builders were found to start their owner-occupied career with a shack made of a mixture of materials as traditional building materials used to construct a typical thatched house are difficult to come by at affordable rates. In such cases, the perceived need to improve shelter produces a fairly adequate form of housing with brick walls, a tiled or sheet roof and a toilet/bathroom unit over time.

This method of incremental building is a well understood means by which slum-dwellers deal with fluctuating and often low incomes and, as such, can be considered to be a survival strategy: namely, 'the ways in which urban poor households organize their production and consumption activities to guarantee the maintenance of their members' (Chant 1991: 26). Although particular strategies are determined by factors such as the size and the regularity of income, the age and conflicting interests of household members, and changes in migrant life, all low-income households have to make choices and set priorities (Papanek and Schwede 1988: 79–80, Mencher 1988: 114). As Turner (1976: 61) put it, 'As long as there are unsatisfied desires for material goods and services people must choose between the cakes they can afford to eat'.

In Hyderabad, however, one aspect of this 'choice' is to place housing low on the list of priorities especially in the short term. Indeed, of 230 slum-dwellers who were asked to list five priorities of expenditure in order of importance, just 6 per cent mentioned housing as one of their priorities and even then it was placed only in fifth place. This is probably an accurate assessment of the availability of the means to undertake housing improvements and reveals a reluctance among low-income households to make long-term financial commitments. Most survival strategies and budgeting patterns of the households included in the sample had short-term planning horizons. It was found that if households take out loans to improve housing, they try to keep the period of indebtedness as short as possible. Nevertheless, the survey found that 53 per cent of all households had some level of debt and that about one-third of these cases were related to housing construction. Indeed, as far as housing finance was concerned, only 20 per cent of the households relied solely on savings and the selling of assets, and 1.4 per cent had received finance from banks or housing finance corporations. The majority obtained finance from the informal sector. Below, we analyse the operation of the three most important sources.

Loans from friends, relatives and neighbours

The terms and conditions of loans from friends, relatives and neighbours vary across time and place, and depend on the urgency of the need for finance and the strength of the personal bond between borrower and lender. Although loan amounts can range from Rs 50 to 30,000, most loans are toward the bottom end of this range. In order to obtain a loan, the borrower relies on non-conventional collateral and creditworthiness is measured in terms of social behaviour rather than in terms of material assets. Often, the loans are given free of interest, but when interest is charged, rates of up to 12 per cent per month have been recorded (above those charged by many moneylenders). The advantage of such borrowing is that conditions for repayment are more flexible and may become open-ended, essentially converting the loan, partly or completely, into a gift. In general, however, there is substantial social pressure to repay the loan according to the initially agreed upon terms and conditions.

Moneylenders

Finance relations with moneylenders, in contrast to those when finance is acquired from friends, relatives and neighbours, are determined by commercial business motives. Full-time moneylenders, whether registered or not, may have a shop or serve their clientele at the doorstep. Part-time moneylenders can be government employees, professionals, or a tenant of the borrower.

The limits of the potential loan depends on the assets owned by the borrowing household, its income, its relationship with the moneylender and whether or not collateral such as utensils and ornaments can be provided. The Hyderabad survey found loan amounts varying from Rs 200 to 30,000 and repayment periods ranging from six months to six years. The interest rates charged by moneylenders depend on the same factors as determine whether a loan is made. Although interest rates vary considerably, compared with bank rates, they are mostly high: in 1993–4 the typical interest rate in Hyderabad was 3–5 per cent per month. Nevertheless, the 'conventional' idea that all moneylenders are squeezing those on low incomes by charging extreme interest rates is not borne out by the findings. To a large extent, the interest rate is determined by opportunity costs and risk premiums and cannot simply be attributed to an assumed monopoly position of moneylenders (Adams 1992: 14–16).

An important cost incurred by moneylenders is enforcement. Mounting debts due to the non-payment of monthly interest charges may ultimately lead to the confiscation of assets by the moneylender. Such practices are often supported by slum leaders, as in the following example from our interviews:

> A coconut-seller cum part-time moneylender was given a loan of Rs 5,000 against an interest rate of 3 per cent a month and it was agreed in writing that the loan would be repaid within one year. If the loan was not repaid on time, the house which was mortgaged could be confiscated. This agreement was sanctioned by the local leader of the settlement.

Most of the moneylenders encountered in Hyderabad were males, who made a number of distinctions between male and female clients. One moneylender said that he preferred to lend to women, because they have a better repayment record and are more polite:

> When gents come they say *namaskar* (Indian way of greeting somebody) and are very friendly in order to get a loan. However, when they come to repay the loan they just throw the money in front of me and ask for the article pledged. They want to show that they are there. Compared with this, ladies stay calm.

Another moneylender said:

> I prefer to do business with ladies. Half of my male clients come for a loan to meet their gambling or drinking habits. Women come for a loan for the hospital, food or school fees. Moreover, ladies do repay the loans better than men do. Ladies come in groups and there are always some known persons in the group. New persons should be brought by a known person.

These comments support the notion that women are seen as better financial managers at the household level, but that in their struggle to make ends meet they are more directly confronted with the consequences of outstanding debts. In Hyderabad, it was found that men tend to prioritize personal expenditure above expenditures for the household. Yet, while it is the women who urge the expedient repayment of loans, it is usually the men who appear to take the main housing finance decisions with or without discussing these with the wife/partner. Obviously, in female-headed households, housing finance decisions are the responsibility of the woman alone, but deprived of other income earners, the opportunities for spreading responsibilities among several members are limited. Thus, while repayment may be prioritized, the means to meet this priority may be lacking.

Chit funds

Chit funds are savings and credit associations and represent a popular informal finance system used by large sections of the Indian population. The chit funds in which slum-dwellers are involved are largely unregistered and can be likened to financial self-help organizations, which are found all over Hyderabad. The most popular form of chit funds among the slum-dwellers is the ROSCA (rotating savings and credit association), in which participants deposit savings at prescribed intervals – daily, weekly or monthly – to a common fund, which is allocated in its entirety to each member in turn. A typical arrangement is for about twelve participants to contribute Rs 100 per month with the total allocated to each member in turn so that in any twelve-month period each participant has received the pot once and the cycle is closed. Sometimes, a deduction is made on the kitty in order to compensate the organizer for their responsibilities.

Two allocation mechanisms can be distinguished: lottery and auction chit funds. In lottery chit funds, the participants contribute a fixed amount per month with the sum of the contributions allocated to one of the participants by means of drawing lots. The probability of an allocation being made to a given participant is determined by chance and should not be subject to manipulation. In an auction chit fund, the participants influence the allocation of the kitty by offering the highest discount on the pot. For example, if the monthly fund is Rs 1,000 the person offering the highest discount, say Rs 300, gets the fund. In this case, he or she takes home Rs 700 with the remaining Rs 300 distributed in equal shares among the remaining participants. If the competition for obtaining the fund is high, such as at the beginning of the ROSCA cycle, the discount will be greater. If, as at the end of the cycle, there is only one member left to obtain the fund then no discount is needed. In other words, demand determines the price of the fund.

Informal housing finance illustrated

Of the households using informal finance, 56 per cent depended on a single source. Apart from savings in kind and money (36 per cent), the main sources were loans from friends, relatives and neighbours (21 per cent), loans from

113

moneylenders (14 per cent) and chit funds (17 per cent). Forty-four per cent of households used a combination of sources as illustrated by the following case studies.

Rasheed and Zaidia

Rasheed and Zaidia have five children (four girls and one boy). Rasheed works as a rickshaw puller but due to health problems earns only Rs 600 per month. Zaidia runs an unregistered shop earning a monthly income of Rs 200. In 1988, they were forced to vacate their rented unit in one of Hyderabad's slums and decided to buy a piece of non-titled land for Rs 6,000. In order to cover these expenses Rasheed participated in an auction chit fund, receiving Rs 4,000 while a further one-year interest-free loan of Rs 2,000 was obtained from Zaidia's brother.

After they had moved on to their plot, a simple one-room house was constructed with scrap and new materials, for which four labourers were hired for two days although part of the work was done by Rasheed and Zaidia. The total cost of the new house was Rs 4,900. In order to generate additional money for construction Zaidia's gold ornaments were sold for Rs 1,000 and Rasheed participated in two more auction chit funds, this time receiving a total of Rs 2,000. Again Zaidia's brother was asked for financial aid and a further Rs 1,000 was given interest-free. However, these sums were insufficient to cover construction costs and Rasheed decided to sell his rickshaw, for which he received Rs 900. From that time onwards Rasheed had to rent a rickshaw to make a living.

In 1992, Rasheed and Zaidia decided to add a boundary wall and a toilet to the house using new materials. Two construction labourers were engaged for a period of five days and the work cost Rs 3,000. This time, Zaidia participated in a lottery chit paying Rs 150 each month for a period of twenty months, eventually receiving the monthly fund.

Guavaya and Radhika

Guavaya and Radhika have four teenage children: a boy and three girls. Guavaya is a letter-setter in a printshop and is the household's sole income earner. He earns a fairly regular and secure income of about Rs 1,000 per month, which is just enough to cover the household's basic costs. In 1984, the family moved to their present settlement due to its proximity to Guavaya's place of work. Initially, they rented a modest unit at low rent but by 1993 rent increases to Rs 200 per month motivated the family to occupy a 12 square metre plot of land. At first they put up a temporary hut, which was demolished soon after by the Public Works Department, but gradually security was obtained and a brick-built house with a thatched roof was put up. House construction involved the purchase of building materials which had to be paid for in a lump sum and the hiring of a skilled mason and two unskilled construction workers at a cost of Rs 370. The total amount needed for the construction of the new house was about Rs 7,000.

114

In order to cover the construction expenses the family used its savings of Rs 1,200 which had been deposited in a bank account and an interest-free loan of Rs 2,000 from a friend who had a secure government job. The repayment schedule was flexible and by 1994 the family had repaid half of its debt. Additionally, with nineteen other women, Radhika entered into an auction chit fund to which Rs 250 was contributed each month for twenty months. The first kitty was allocated to the fund organizer as payment for fund management. Radhika received small shares in the fund's monthly balance plus Rs 3,800 when her turn came. All sums were added to the household's housing fund.

Razia Begum
Razia Begum lives with her two teenage children (one boy and one girl). In order to make a living she works as a housemaid and her son of 16 is employed as a handyman at a watch repair shop – total household income is Rs 600 per month. In 1977, a 20 square metre plot of land was bought from a local leader and a hut was constructed. The total costs were Rs 900, which was a gift from Razia's sister. However, the hut, especially the roof, required regular repairs and, in 1992, the last repair on the roof, using bamboo sticks, mats, plastic and *danbar* sheets, cost Rs 1,500. To pay this cost Razia obtained Rs 500 as a gift and Rs 1,000 as an interest-free loan from her employer. The loan was repaid by means of a deduction of Rs 50 from her monthly income.

On the basis of the above discussion the general characteristics of informal housing finance can be summarized. First, the amounts spent on housing are relatively low and consequently the houses constructed are modest: Guavaya spent Rs 7,000, which was equivalent to seven months' household income; Rasheed and Zaidia spent Rs 13,900 on both land and housing, the equivalent to seventeen months' household income; and, Razija spent Rs 1,500 or just over two months' household income on the last repair of her hut.[7] Second, it is clear that housing investment takes place when it is considered necessary and convenient by the dwellers and that financial obligations are kept to the short or medium term. They are also spread over several sources to minimize the chances of getting caught in a debt trap.

The third characteristic concerns the cost of informal housing finance. Obviously, interest-free loans are preferred, but, even when available, they are rarely sufficient to cover construction expenses. More usually, informal housing finance can be expensive on account of the interest charged, especially when borrowing comes through a moneylender. However, even when loans are made via friends, relatives and neighbours, interest can be high. In the case of auction chit funds, on the whole there is a link between the price and the stage of the ROSCA cycle. In this respect it should be noted that the affordability of informal housing finance is related both to the charges imposed and to the social conditions of the loans. In contrast to other sources of informal housing finance, chit funds combine the mobilization of savings among participants with the provision of finance.

Fourth is the importance of gender and social vulnerability. Many lottery chit funds involve women both as operators and participants while in auction chits most contributors are male and the managers are almost always men. Informal finance, therefore, should not be universally regarded as socially progressive. Marginal and vulnerable households, a disproportionate number of whom are female-headed, face more difficulties in getting access to loans than the somewhat better-off among the low-income population. In addition to the assets and savings of potential borrowers, informal lenders require evidence of the level and security of income, repayment records covering the recent past, or ownership of a house or a plot as a precondition to granting a loan. All these requirements work against women.

Fifth, there is clearly a two-way enabling relationship between informal finance and housing. The deposit of savings on a compulsory basis in a chit fund, for example, can be regarded as a form of collateral by other (informal) agents. As a related point, informal finance attaches importance to networking as loans will be given only if the borrower can be trusted and, therefore, the borrower's character and behaviour should be known by the lender. This is straightforward when it concerns relatives, friends, colleagues or employers, but in the chit funds this notion of trust is the responsibility of either the group or organizer. A particular problem for low-income households is the difficulty of establishing trust with strangers by means of entertainment due to their limited incomes and consequently they face difficulties maintaining a network to enable access to finance.

CONCLUSION

The majority of the EWS households in the cities of Andhra Pradesh attach a low priority to housing improvements in the short term. If they improve their house at all, they do so in an incremental manner and use various sources of informal housing finance. If such an incremental process is affordable it is not because loans are necessarily cheap, but because the borrowers themselves decide on the timing and form of investment, use relatively small loans and keep repayment schedules flexible and short to medium term. Creditworthiness is mainly or completely measured against a non-conventional collateral.

By intervening in a relatively large-scale manner, the public sector believes that it can make things better. If one exclusively focuses on the ultimate physical outcome of the EWS housing programme, the production of permanent houses, the results are indeed impressive. Moreover, at first sight, the loan conditions seem favourable: interest rates are low and repayment is spread over a long period, allowing for relatively modest monthly instalments. However, as illustrated by the above description of the prevailing construction and finance patterns among the low-income population, such loan conditions interfere with EWS survival strategies. A serious shortcoming of the WSHP is that the public loan-cum-grant only covers roughly 40 per cent of the real unit costs. As a result a large number of beneficiaries are forced to obtain additional informal finance which, combined with the required investment of a large share of income in the programme house and temporary shelter, draws many into a debt trap. A significant proportion of the EWS households who have been able to stay in the programme houses have lost a substantial part of their disposable income on the repayment of their informal loans. At best they have gone through a prolonged period of severe hardship while, in the worst cases, they have been incapable of gradually improving their financial situation and remain trapped by the scale of their debts. As a result, many may have lost the opportunity to invest in a petty business, a job or in the education of their children.

Ultimately, many of the so-called beneficiaries are forced to sell or rent their houses. Among the displaced, it is female-headed households and those with younger children that are among the first to be pushed out. In their concern with household budgets, women are most directly affected by the consequences of the programme. They are confronted with the task of feeding and clothing their families within the confines of a tightened budget over a long period of time. In addition, the acute financial problems may have led many women to sell their gold ornaments which are not only a source of pride and dignity, but also a form of personal insurance against unforeseen events such as divorce or the death of the husband. Without this form of personal insurance women are prone to further risks.

Given that improving shelter is not a high priority among many low-income households and given that research findings suggest that housing finance programmes exacerbate financial hardship while trying to resolve the problem, one might pose the question as to whether housing finance should remain part of government policy. Although housing officials are aware of the shortcomings of the finance programme to date, they tend to blame the beneficiaries for selling their houses. But, if the government of Andhra Pradesh wants to continue its housing finance tradition, it might adopt the more flexible arrangements which low-income people themselves use to finance improvements. Such arrangements may not lead to the instantaneous 'eradication' of huts and the widespread emergence of 'decent' permanent structures, but it might offer finance which fits the survival strategies of low-income people and which allows for a modest and incremental improvement of the EWS housing stock.

NOTES

1 Our thanks are due to Wiebe Nauta, Jan van der Linden, Joost van Loon and the editors for their insightful comments on earlier drafts of this chapter. We would like to thank

the Dutch Foundation for the Advancement of Tropical Research (WOTRO), the Dutch Ministry for Development Affairs, and the Vrije Universiteit, Amsterdam for funding the research on which this chapter is based

2 One Indian Rupee = 0.033 US $ (1993–4).

3 The sliding scale establishes an interest rate of 12 per cent on loans of Rs 25,000, 15.5 per cent on loans of Rs 25,001–100,000 and market rates on loans above Rs 100,000.

4 Approximately 31 per cent of the households in squatter settlements had a monthly income of Rs 1,000 or less, 28 per cent earned more than Rs 2,000, and 67 per cent of the principal income earners were engaged in low-paid jobs. By contrast, in scheme settlements 19 per cent of households had an income of Rs 1,000, 54 per cent earned over Rs 2,000 and 55 per cent worked in low-paid jobs.

5 According to author calculations, the minimum monthly income for a four-member household would be Rs 730–860, depending on possession of a ration card which entitles the holder to certain basic goods at a subsidized price, but excluding the cost of clothing and education for the children, and unforeseen events such as illness and occasional expenses such as marriage/dowry.

6 This section is based to a large extent on Smets (1996b).

7 These calculations are based on household incomes at the time of the survey and may have been higher or lower at the time of investment.

8

AFFORDING A HOME

The strategies of self-help builders in Colombia

Katherine V. Gough
University of Copenhagen, Denmark

INTRODUCTION

It is now widely accepted that the vast majority of inhabitants of cities in developing countries have to be involved in the construction of their houses if they are to become home-owners. During the construction process, most households will make payments for land, construction materials and skilled labour. This chapter analyses the different ways in which households finance these payments and focuses on attempts by self-help builders, by community-based initiatives and by various governmental and non-governmental organizations, to reduce the cost of constructing a home. The chapter argues that housing strategies vary according to a household's resources and networks, and that the housing process itself offers implicit opportunities for maintaining financial control.[1] The final section indicates how most community and institutional-level strategies encounter difficulties as they are inflexible to changes in household resources and the construction process.

The chapter is based on fieldwork conducted in the intermediate-sized Colombian city of Pereira.[2] Research consisted of a questionnaire survey with 243 randomly selected households in five low-income settlements, in-depth interviews with twenty households, and participant observation.[3] The five low-income settlements in which research was carried out were of varying legality and age (Table 8.1). The two most recent settlements include a public sites-and-services project known as '2,500 lotes', where households were in the early stages of construction, and Villa Santana which is a pirate (illegal) settlement and where the houses were not more than five years old at the time of initial survey work. The three older settlements were between eight and ten years old: El Plumón originated as an invasion along a disused railway line; Leningrado was founded by a self-help housing association and Las Mercedes was a sites-and-services project.

Table 8.1 The study settlements

	2,500 lotes	Villa Santana	Leningrado	El Plumón	Las Mercedes
Origin	Sites-and-service	Pirate urbanization	Housing association	Invasion	Sites-and-service
Legal status					
a. at foundation	legal	illegal	illegal	illegal	legal
b. when studied	legal	being legalized	legal	illegal	legal
Age when studied	1 year	5	10	10	8

The two sites-and-services projects, 2,500 lotes and Las Mercedes, were legal and had services installed from their inception. All the houses in Las Mercedes had been connected to the mains sewerage, water and electricity supply but in 2,500 lotes not all the households could afford to make the necessary connection to the mains: 16 per cent were using a neighbour's toilet, one-third had an outdoor tap in their plot, and a few had made an illegal connection to the electricity supply. El Plumón, Leningrado and Villa Santana all originated as illegal settlements but Leningrado had subsequently been legalized, a legalization programme for Villa Santana was under way and one was being discussed for El Plumón.[4] All the houses in El Plumón and Leningrado had electricity, water and sewerage, but in Villa Santana 18 per cent of households had only an outdoor tap in their plot and 17 per cent were not connected to the sewerage system.

The standard of the houses in 2,500 lotes and Villa Santana was very similar: a little over half of the houses were built of temporary materials (usually bamboo), some of which were of very poor quality. In the older sub-communities, few houses were built out of temporary materials and many were well consolidated with most households having improved the quality of the construction materials and the size of the houses over time. The ability to consolidate was not ubiquitous, however, and some households were still occupying a very small dwelling even after ten years (Gough 1996a). Two particular features stand out. First, it appears that households in the latter stages of consolidation are better-off than those in the early stages. Thus, while over half of the households earned one minimum salary or less, the mean household income was higher in the older settlements. Second, household structure varied with settlement and housing conditions. Of the households interviewed, 76 per cent had both male and female heads, 20 per cent were female-headed and 4 per cent were male-headed. Over half of the households had between four and six members, few had more than ten members. Overall, 70 per cent of households were nuclear and 30 per cent were extended. Extended households were more common in the older settlements, where the average household size was slightly larger than in the younger settlements (Gough 1992).[5]

The chapter presents the experiences of four households, one from 2,500 lotes, two from Villa Santana and one from Leningrado, to illustrate various strategies to finance housing construction. The Ortega household lived in 2,500 lotes and when first interviewed consisted of a widow in her early thirties with three young sons; they were subsequently joined by Sra Ortega's new partner. Sra Ortega had to work in a bar in the evenings to support her sons while single, but after her partner (who was a construction worker) moved in she gave up work. The Ramírez household consisted of a woman in her late fifties who lived with her teenage daughter in Villa Santana. Sra Ramírez worked one day a week for a lottery agency but her daughter was unable to work as she suffered from epilepsy. Their income was supplemented by begging and help from neighbours. Also in Villa Santana was the Carrillo household, a nuclear household. Sra Carrillo was a trained nurse and her husband a construction worker. They were both in their thirties and had three young children. Finally, the Arango household was an extended household: Sr and Sra Arango were in their fifties and lived with four of their eight children, plus two grandchildren and a niece. The two daughters living at home worked as domestic servants and contributed to the household income but the two sons were unemployed. Sra Arango was a housewife and Sr Arango only had temporary manual work. To appreciate why these households had to rely on their own resources and financial strategies it is useful to provide a brief overview of the development of housing finance in Colombia.

PUBLIC-SECTOR HOUSING FINANCE IN COLOMBIA

Since the 1930s, Colombia has undergone a transition from a rural society to a predominantly urban one. The proportion of the population living in urban centres increased from 29 per cent in 1940 to 70 per cent in 1990 (Wilkie *et al.* 1990). Despite the establishment of the Instituto de Crédito Territorial (ICT) with specific responsibility for government housing programmes, housing was a low priority in national development policy (Betancur 1995, Gilbert 1997).[6] Prior to the 1950s, the government had virtually no policy on housing production for low-income groups and it was not until the early 1960s that public-sector construction programmes were established on a significant scale (Gilbert and Ward 1982). Thereafter a declining flow of funds limited public-sector construction and the aimed-for production was never realized (Stevenson 1979). The failure of government production was followed by the 1970–4 and 1982–6 development plans, which respectively sought to promote the construction industry as the leading sector of the economy and to increase housing construction for the poor. However, the cheapest housing built by the private construction industry was too expensive for low-income households so that about one-third were left to find rental accommodation and about one-half to provide themselves with their own homes through self-help construction, the majority in pirate urbanizations and a few in invasions (Gilbert 1997).

In recent years the Colombian government has re-oriented its housing policy in line with World Bank policy to abandon the government's role as a producer of housing and to reform policies, institutions and regulations to enable housing markets to work more efficiently (World Bank 1993). There are now fewer controls on housing finance, and a system of housing subsidies aimed at low-income communities distributed through local communities, NGOs and municipal governments has been introduced. These direct subsidies can be used to purchase a serviceable plot or a core housing unit, to make home improvements, pay for land titling and service acquisition. The government, however, has not been able to afford to meet all of the subsidy requests and more than half of the households who have been granted subsidies have yet to receive them (*El Tiempo* 16 June 1997). Gilbert (1997) claims that although emphasis was placed upon helping the poorest households, many have been unable to take up their subsidy because they could not afford one of the available housing solutions or to pay the deposit. As before, therefore, most low-income households in Colombia have been obliged to self-finance housing construction.

FINANCING A SELF-HELP HOUSE

To build a self-help house most households will have to purchase land and construction materials as well as hire labour (Table 8.2). In Pereira, only 5 per cent of those interviewed had not paid for the land, 8 per cent had not purchased any construction materials, and 27 per cent had not hired any skilled labourers. The percentage of households who had paid for labour and/or construction materials was higher in the older settlements, suggesting that as houses are consolidated there is an increasing tendency to hire labour and to buy construction materials.

Table 8.2 Percentage of self-help builders involved in financial transactions during house construction

	2,500 lotes	Villa Santana	Leningrado	El Plumón	Las Mercedes	All Settlements
Bought plot	100	94	100	83	100	95
Bought construction materials	90	90	97	100	100	92
Hired labour	46	60	71	65	79	63

Households raised money to finance land purchase and house construction in a variety of ways (Table 8.3).[7] Household savings were the most frequent source of funds. In order to generate surplus income which could be spent on housing, household members often worked extra hours, took on an additional job or took up paid employment. As has also been found to be the case under conditions of increasing economic hardship in Mexico, it was especially common for women to increase their income-earning capacity, often entering the labour market for the first time since marrying, while others attempted to diversify their income earning opportunities (González de la Rocha 1994).

Another strategy used by about 20 per cent of the households was to establish a home-based enterprise as a means of earning extra income: most common are the small-scale sale of food and dressmaking, both predominantly carried out by women (Gough 1996a). Money earned from home-based enterprises has clearly helped households finance the consolidation of their homes; in turn, as the houses have been consolidated, the households have had greater possibilities for setting up a home-based income-generating activity whether it be a store, workshop or the renting out of rooms. The predominant role for women in this process illustrates the way in which 'cultural definitions of women as wives and mothers intervene in economic processes' (Bohman 1984: 129). Women try to find economic activities which are compatible with their roles as wife and mother. Especially for women without male partners, working in the home provides the only source of income which they can combine with their domestic responsibilities (Gough 1996a).

According to Boleat (1987), of the four main types of housing finance system (direct, contractual, deposit taking and mortgage bank), the most common in developing countries is the direct route through which funds are obtained directly from those who have them (Rueda *et al.* 1979, Strassman 1982, Macoloo 1989). This would also appear to be the case in Pereira, where some households financed the construction of their houses by gaining a lump sum by winning a lottery, being given money by a politician or family and

Table 8.3 Methods of financing land purchase and house construction

	2,500 lotes	Villa Santana	Leningrado	El Plumón	Las Mercedes	All settlements
Savings	80	64	70	92	66	71
Lump sum	38	42	30	33	38	38
Liquidación	18	30	33	33	28	26
Loan	16	7	17	25	38	17
Sold Possession	—	3	7	—	—	2
Sold/swapped land/property	—	11	7	8	13	8
Sample size	50	74	30	12	32	198

Note: Percentage totals are greater than 100 since households combined several methods of raising finance.

friends, or from an inheritance. Household members who worked in the formal sector sometimes took out one or more retirement payments (*liquidaciones*) from their place of work. A few households raised money by selling possessions, a plot of land or a house. The acquisition of loans was not very common. Banks and other financial institutions were not willing to grant self-help builders loans and the irregular incomes of many households made them wary of having a loan and the commitment to making monthly repayments. Overall, only 17 per cent of households took out a loan and of this number 35 per cent were obtained from the workplace and another 30 per cent from family and friends.

An important source of finance was the Instituto de Crédito Territorial (ICT), which provided loans to 35 per cent of households in the public-sector projects. In 2,500 lotes, in order to qualify for a loan, households had to have completed their house foundations, and the loans were tied to the purchase of construction materials and the hiring of labour. The materials had to be purchased from the construction materials centre (see below) but there was no restriction on what type of construction materials could be bought. By making the centre the sole supplier of construction materials bought using loans, however, self-help builders were prevented from buying directly from the source of their choice. For households, the most common complaint was delays in loan allocation while for the ICT the main problem was the recovery of loan repayments since no strict means testing had been done in order to avoid discriminating against people working in the informal sector.

For some households, the loans did speed up the rate of consolidation and hence reduced either the need or the length of time that households had to live in temporary bamboo dwellings. The loans also enabled the purchase of materials in larger quantities and thus provided economies of scale. However, despite the availability of ICT loans in Las Mercedes, the standard of housing was no higher than in the other communities of a similar age. Over time, many of those households who had not received a loan managed to catch up to the standard of housing of those who had (Gough 1992).

Many households combined different strategies for obtaining finance but there was a predominance of informal methods for raising funds for self-help construction. This was the case for all four case study households, though the actual method of raising finance varied according to the household's circumstances. Initially, Sra Ortega financed the construction of a bamboo house from her income, but when she was joined by her partner they financed the construction of a brick house by using money from an inheritance and a construction materials loan from the ICT. The transition from being a female-headed household to a joint-headed household greatly increased Sra Ortega's financial resources. Sra Ramírez did not make any financial outlay for her house as, due to her critical financial situation, both the land and materials were received as a gift. Sra Carrillo's income-earning capacity as a nurse was a vital source of extra income for the construction of the house and gave her the opportunity to take out a retirement payment (*liquidación*) from the hospital. After the house was completed well enough for occupation, Sra Carrillo stopped working as a nurse and stayed at home to look after the children. As an extended family, there were several adult members of the Arango household who were able to pay for the construction of the house from income (savings) and the acquisition of the plot in monthly instalments to the self-help housing association Provivienda. These cases show that a range of informal methods for obtaining finance are used by self-help builders which vary according to the nature of the household, the employment of household members, the resources of the household and the settlement type. The strategies used by self-help builders to reduce the costs of house construction will be discussed below.

STRATEGIES TO REDUCE THE COST OF SELF-HELP CONSTRUCTION

Self-help construction is, by its very nature, an individual process which is strongly influenced by a household's resources and priorities. Here, the ways in which households attempt to reduce the cost of self-help construction are discussed by looking at the methods used to obtain a plot of land, the system of house construction, labour strategies and the types of construction materials used.

Obtaining a plot of land

The first obstacle to overcome in self-help house construction is finding a plot of land. Most households when asked why they bought a particular plot of land emphasized that there had been very little choice. They bought whatever land was available, regardless of whether it was in a sites-and-services project, a pirate urbanization or an invasion. Obviously invading land is one method of reducing costs, but success is highly dependent upon the degree of tolerance by the local authorities and most households interviewed in the invasion of El Plumón had bought land from the original invaders. Most households in a pirate urbanization, therefore, had to purchase a plot.

More than half of the households interviewed had found out about the availability of a plot of land for purchase or invasion via family and friends. Others had found their plot by looking and asking, through local politicians or by applying for a plot in a government or self-help housing association project. Sra Ortega was allocated a plot of land in 2,500 lotes after the house in which she had been renting a room was demolished for safety reasons. Payment for the plot was made on a monthly basis to the ICT. Sra Ramírez obtained her plot of land in Villa Santana after having regularly attended meetings held by the politician and pirate urbanizer who established the settlement. She was given the plot of land as she did not have the means to pay for it. The Carrillo household bought their plot of land in Villa Santana from a friend to whom they paid a lump sum and the Arango household obtained a plot of land through the self-help housing organization, Provivienda, which they paid for in monthly instalments.

The case studies illustrate the importance of the economic and housing circumstances of households for the allocation of plots of land, and the significance of being affiliated to a political group. Sra Ramírez and the Arangos were allocated land through their affiliation to a politician/political group. The poor economic and housing circumstances of Sra Ramírez and Sra Ortega resulted in the former not having to pay for her plot and the latter being allocated a plot in a sites-and-services programme. The Carrillo's financial circumstances enabled them to buy a plot in Villa Santana, albeit at a favourable rate as they bought the plot from friends.

System of construction

When households have obtained a plot of land, they face the problem of having to bear the costs of construction while also having to pay for their existing accommodation elsewhere. Two basic strategies are commonly adopted by self-help builders to reduce this double burden. First, to quickly construct a shack of temporary materials or a room in brick for immediate occupation, and second, to live with relatives (rent-free) while working on a larger, more permanent structure. The latter strategy was adopted by households who either considered construction from temporary materials to be a waste of money or were unprepared to live in a shack. The Carrillo family, for example, lived with relatives for three years while they constructed their single-storey house built of brick. The majority of households in Pereira, however, constructed some form of initial housing and discussion will be limited to this strategy.

Overall, 61 per cent of the households interviewed had moved into a dwelling within one month of starting to build. Two-thirds of households started by constructing a shack of temporary materials with walls of split bamboo (*esterilla*) which was sometimes covered with plastic sheets in an attempt to keep out the rain, and with a roof of second-hand clay tiles or bitumized cardboard (*cartón*). The ratio varied between the settlements: 92 per cent of households in El Plumón started with a bamboo shack compared with only 47 per cent in Las Mercedes. These variations can, at least partly, be explained by differences in the legality of land tenure at the time of the foundation of a settlement. Where land was invaded, households had to construct rapidly in order to lay claim to the land and were reluctant to build using permanent materials until there seemed no likelihood of eviction. In government projects, households possessed legal rights to their plot from the start and in pirate settlements the households felt secure about their land rights since the plots had been purchased. Households in invasions, therefore, were much more likely to start by constructing a temporary shelter than households who had purchased their land.[8]

The aim of all households living in temporary structures was to consolidate their dwelling and the majority of them succeeded in doing so. The houses were gradually consolidated in different ways as and when the household could afford to make some improvements. Some households built extra rooms in bamboo or replaced existing materials with those of better-quality, but still temporary, materials. Sometimes brick walls were built around the outside of the bamboo walls and when the bamboo house was completely surrounded it was taken down. Sra Ortega, for example, originally built a bamboo shack but having been joined by her new partner built a brick house on the plot. While the building was in progress they lived in a neighbour's house which was not yet occupied.

An alternative method was to build a brick house in stages, one room at a time, and usually with an intermediate stage where the front part of the house was in brick with the back still in bamboo. The Arango household initially built a one-room bamboo shack and after three years laid foundations on half of the plot and started to build in brick. When interviewed, they had two brick rooms at the front of the house and the original bamboo room at the back was used as the kitchen. Some houses were built progressively of brick from the start with households erecting one brick room in which they lived while slowly constructing the rest of the house at the back. Whatever strategy was adopted, the aim was to reduce the cost burden of construction.

These cost-cutting approaches do possess a number of significant short-comings. First, cutting costs means that not all households have succeeded in consolidating their dwelling. The Ramírez household was still living in a tiny bamboo shack five years after they had built it. Second, for many house-holds, the initial self-built house was small and of poor quality materials, and was often of a lower standard than the household's previous accommodation. Third, the temporary construction and disruption surrounding self-help particularly affected women as they spent many hours in the house and had the responsibility of caring for the children and cooking in very cramped conditions. Children were also affected; one woman interviewed was very distressed because her 10-year-old son had left the family for the streets of Pereira shortly after moving into a bamboo shack in 2,500 lotes. Her son claimed to prefer the freedom of the streets to being trapped in such a small house. His mother had tried to bring him home on several occasions but each time he returned to the streets. She feared that by the time they had managed to extend the house, he would be so accustomed to living on the streets that he might never be able to live with them again.

Labour strategies

Self-help housing is often thought of as synonymous with self-build even though research has revealed consistently that many households employ wage labour in the construction process. Rodell (1983: 30), for example, has argued that paid labour has increased to such an extent that a situation has resulted where 'family labour does not, as initially thought, play a dominant role in self-help' and Macoloo (1989: 188) has claimed that in Kenya 'self-build as a form of housing production clearly does not exist any more in irregular settlements'. In Pereira, however, there is evidence that the unpaid labour of households, family and friends remains an important strategy for reducing the cost of a self-help house.

Overall, 75 per cent of households participated at some point in the construction of their homes and 11 per cent claimed that they had constructed their dwelling entirely with their own labour. The presence of male household members who were construction workers, such as Sr Carrillo, greatly reduced the need to hire labour. The Arango household, for example, was slowly constructing their house without any outside help. They were able to do this as theirs was an extended household with several male adults. More often, household members worked alongside a hired builder performing the unskilled tasks as instructed and thereby saving the expense of employing building assistants. When Sra Ortega built her bamboo house she and her young sons worked alongside a paid labourer while their brick house was built by her new partner who was a construction worker by trade.

The contribution of the labour of household members tended to be underestimated during interviews and was only revealed through careful probing. A wide range of household members participated in the construction process, from young children to old women. In particular, women's participation was underestimated, especially by male household members, despite women clearly being active in the construction of their own homes (Moser and Chant 1985, Nimpuno-Parente 1987, Peake 1987, Vance 1987). As a result of their involvement, many women were very knowledgeable about the construction of their houses and when both men and women were present, it was often the woman who could remember more clearly details about the construction process. However, although women participated in construction, female-headed households, without any adult male members, were unable to build their houses themselves without some external (paid or unpaid) help.

The unpaid help of unskilled family and friends was also an important aspect of self-help house construction: 43 per cent of households reported that they received some unpaid help from family and friends and 37 per cent of households had built their dwelling without any recourse to paid labour. These helpers were of all ages and abilities and of both sexes. Some were neighbours, while others were friends or family living in other parts of the city. The Ramírez household, for example, had built their bamboo shack with the unpaid help of a nephew. Unpaid help was particularly important in the early stages of construction when budgets were stretched and the tasks less skilled. There was a trend, however, whereby self-help builders were increasingly having to pay for the skilled labour of family and friends as economic hardship meant that many construction workers had to charge for their time.

Construction materials

The image of self-help builders constructing their houses entirely out of scavenged materials has long been dispelled. Construction materials have been found to account for 60–65 per cent of the non-land cost of housing in general (Lowder 1986) and for 86 per cent of the cost of self-help housing on average in Colombia (Rueda *et al.* 1979, Gough 1992). It is most common for self-help builders to purchase construction materials in small quantities mainly from nearby retailers where prices tend to be high (Gough 1996b). However, self-help builders adopt a range of strategies to reduce the cost of the materials. Frequently, builders bought materials which were of poor quality or were second-hand, especially clay tiles, bricks and wood. Some of these materials were purchased from small-scale, local construction materials merchants who catered specifically for the needs of self-help builders. Second-hand materials were also purchased from other areas of the city where houses were being demolished. The Carrillo family, for example, built their roof of clay tiles which Sr Carrillo had obtained cheaply having been employed in demolition work and other materials were bought second-hand from local construction materials merchants.

It was also common for households to be given some construction materials: the most frequent gifts were wood, clay tiles, *cartón* and bamboo. One-third of all households surveyed had been given at least some construction materials, and 8 per cent had been given all the materials in their house. The donors were usually family, friends or neighbours, but included politicians. Common scenarios included being given wooden doors and windows from parents, bamboo from friends who had dismantled their shack after constructing a brick house, and donations of materials from politicians either to very poor families or to a community for general distribution. As her financial situation was so dire Sra Ramírez was given not only a plot of land, but also bamboo, *cartón*, and wood with which to build a shack by the politician who established the settlement. The Arango household had also been given many of their construction materials by relatives. Gifts of materials were most common in 2,500 lotes and Villa Santana, suggesting that households are more likely to receive donated materials in the early stages of construction when they are building temporary accommodation and thereafter are required to purchase materials for more permanent structures.

While no household actually admitted to stealing materials in order to reduce the cost of construction, self-help builders often reported the theft of some of their materials, especially those which had to be kept outside the house, such as sand, gravel and bricks. The fear of theft deterred some households from buying materials in large quantities. Construction materials were also stolen from construction sites in other parts of the city.

Most households used a combination of ways to obtain construction materials. Sra Ortega, for example, bought half of the bamboo poles and *esterilla* for her shack from a small construction materials retailer in 2,500 lotes and half were given to her. Clay tiles for the roof were bought second-hand from a merchant in 2,500 lotes and *cartón* to cover the rest of the roof was a gift from friends, as was the wooden door (there were no windows). The concrete floor required seven bags of cement of which four were given by a neighbour and three were bought. Other neighbours sold Sra Ortega sand by the bucket at reduced rates.

Community strategies

It has been argued that self-help builders should work together to reduce construction costs (Turner 1988). One possibility is for a group to buy materials in bulk or establish co-operatives of self-help builders in order to avoid the purchase of small quantities of materials from local distributors (Rueda *et al.* 1979). In 2,500 lotes, the self-help builders were divided into twenty sub-communities with the aim that each would purchase construction materials in bulk and collectively construct houses. A few communities purchased some construction materials collectively, most commonly for the house foundations, but many of these attempts ran into difficulties. In some cases money disappeared after collection or was ill-used. Also, it was difficult for some materials to be distributed equally (such as sand and gravel) and after distribution there was the problem of theft due to a lack of storage facilities. There was also the fundamental problem of people constructing at different rates and hence requiring materials at different times. Furthermore, some households could not afford to buy in bulk even at the reduced rates; poorer households (among which female-headed households tend to be over-represented) were not able to participate in these schemes. Hence, collective and community organization which appears to be economically sound on paper has proven to be very difficult to put into practice (Gough 1996c).

Communal labour

A second strategy based on community organization was to encourage construction by collective rather than individual effort so as to enable the sharing of knowledge and labour power and reduce expensive errors. In 2,500 lotes, some communities had elected to build their houses collectively and communal labour had been used successfully to construct the foundations: probably because the laying of foundations is labour intensive and can be completed relatively quickly. However, attempts to use communal labour for subsequent stages often floundered. One of the problems faced by communities in 2,500 lotes was their size. With many communities having over a hundred members, numbers were too large to successfully co-ordinate communal construction. The cause of this problem in Pereira stemmed from the use of government housing projects for partisan political purposes, which meant politicians who were influential in the establishment of settlements would demand a certain number of plots to be distributed among their followers (Burgess 1986). As a result, a number of communities were far larger than originally intended.

Even so, had politicians not been present and had the communities been smaller, communal work would still have had to overcome the lack of trust among households whose only common bond was the desire to own a house. The heterogeneous nature of self-help housing meant that co-ordinating construction between households with widely varying financial resources and sources of labour and construction materials was impossible. Female-headed households, especially those with young children, found it difficult to participate in the communal labour sessions. Many of the attempts to co-ordinate households in collective construction floundered, therefore, because they were, in effect, trying to regularize what is by its nature an individual process.

Community finance initiatives: lotteries

Some communities in 2,500 lotes tried to establish lotteries as a way to help at least some households to reduce construction costs. In most cases, construction materials such as roofing sheets of asbestos cement were donated to the lottery by a politician or political group. Community members would then buy tickets and the money would go toward the purchase of construction materials to be used in the next lottery. For the household fortunate enough to win, the benefits of the lottery are self evident, but for those which did not it could be argued that the lottery actually increased the cost of construction. Moreover, problems arose in one of the communities when allegations of lottery rigging were made after the community leader won the first and only lottery.

INSTITUTIONAL STRATEGIES

A range of institutions have been involved in attempts to influence self-help construction ever since the Colombian government began actively to support self-help as a method of construction during the 1970s (Betancur 1995). This section looks at the role of various institutions in providing education on the self-help construction process, in supporting the production of materials by self-help builders and in the establishment of a construction materials centre.

Education

Education in construction methods could help to reduce errors and hence the costs of construction. Most of the builders interviewed in Pereira were constructing a house for the first time with the result that many costly and time-consuming mistakes were made in the process (Gough 1998).

131

In order to improve the technical proficiency of self-help builders and the advantages of building collectively, the government organization Servicio Nacional de Aprendizaje (SENA) ran courses in the centre of Pereira, usually one evening a week for those who were constructing in government projects. The teaching format was a typical classroom with an instructor who explained the basics of self-help construction with those present having little chance to ask questions. A range of pamphlets were produced by SENA which described and illustrated the different construction techniques. Although this was well intentioned, many builders had difficulties relating the pamphlets to practice and found the language too technical, especially as many of the builders were illiterate.[9] Despite women's involvement in self-help construction, it was men who were featured in all of the illustrations in the booklets. For others, with little or no education, the classroom format was inappropriate. It was particularly difficult for women to find time to attend the courses and they also tended to be among the least educated.[10]

Production of construction materials

The cost of construction materials can be reduced by self-help builders if they produce the materials themselves and hence eliminate the wage labour component and middlemen. Within 2,500 lotes, two sub-communities received support from outside institutions to make their own materials. The sub-community of Byron Gaviria was assisted by the national self-help housing association Fundación Servicio de Vivienda Popular (SERVIVIENDA) to set up a prefabricated system of cement-based blocks, tiles and slabs, as well as help to get a loan to make bulk purchases of the raw materials. About three-quarters of the fifty households in the community elected to participate in the project. The tiles were made during the week by a paid member of the community and his son, and the blocks and slabs were made by the rest of the community on Sundays. The houses were collectively built by household members. It took just three months to make the materials and construct all of the houses, a significantly faster rate than in any of the other communities in 2,500 lotes and at a lower cost.

One of the main disadvantages was that the prefabricated materials were not strong enough to support a second floor without demolishing the existing house and starting again. Ten years after the prefabricated houses had been built the vast majority had been demolished to enable the construction of larger houses and all the roofing tiles had been replaced due to a faulty design. Moreover, only those households with stable incomes and manageable debts could participate and all of the households interviewed who were participating in the project had at least two incomes.

In the sub-community of Sinai a different approach was used to reduce construction materials costs. Here, SENA lent two machines to produce earth-cement mix blocks. The advantages of the blocks were that they were cheap and could be produced on site, thus eliminating transport costs (Santana and Casabueñas 1981). However, partly due to poor leadership, production of the blocks was very slow and the community had not started to build houses one year after plots had been distributed. Moreover, the use of self-made materials involved disguised costs, lower than expected quality and labour intensive production which left households with less free time than had been expected or the option to hire labour and remove any cost advantage (Bamberger 1982). As was the case with communal labour, it was particularly difficult for female-headed households with young children to participate in the production of the materials.

Construction materials centre

In several Latin American countries, government and NGOs have been instrumental in the establishment of special distribution centres for construction materials located in low-income settlements in order to reduce construction costs. The centres provide space for construction materials producers to sell directly to self-help builders and so eliminate the costs incurred by intermediary distribution and transport agents. In 2,500 lotes, a construction materials centre was established organized by the Fondo de Vivienda Popular Pereira (FVPP) in collaboration with the Fundación para el Desarrollo de Risaralda and Fundación Carvajal de Calí. The construction materials centre played an important role in processing construction materials loans to self-help builders and materials producers were provided with rent free space in return for a 3–7 per cent cut of their takings.

Business in the centre, however, was slow despite a dynamic project leader who went around the community distributing leaflets giving the prices of some of the materials. The centre faced several fundamental problems. First, there was no real incentive for producers to supply the centre directly since they already had an efficient distribution network for their construction materials with formal and informal retailers. Thus, by selling directly to households, the producers were not expanding their market but were just reaching it in a different way. Second, the producers were unwilling to charge low prices to the centre so there was little cost advantage compared to construction materials retailers. Third, construction materials merchants objected to the centre, which had the potential to destroy the livelihoods of the smaller materials retailers who were located nearby. The larger wholesale construction materials merchants also objected to being bypassed and in some cases prevented producers from supplying the centre: no iron was sold in one centre as an agreement could not be reached between the producers and the distributor. Finally, the main construction materials centre was also unpopular among the self-help builders themselves. Not only was there very little cost incentive to purchase from the centre, but the buying procedure was very formal and the centre was a particularly impersonal place. There was no possibility for

chatting with friends and receiving a friendly bit of advice, nor were alle-
giances likely to develop or finance be offered.

CONCLUSION

Most self-help builders are involved in financial transactions in order to construct their houses and most of them raise the necessary finance through direct
or informal means. As illustrated by the case studies, a range of informal
methods for obtaining finance are used and these vary according to the nature
of the household, the employment of household members, their resources and
networks, and the settlement type. The use of household savings was particularly common with members often working extra hours or taking on an
additional job in order to generate surplus income. Households with members
working in the formal sector had the possibility of taking out retirement payments (*liquidaciones*) from their place of work.

Loans were not a very common way of financing construction, with the
exception of households constructing in sites-and-service settlements who had
access to government loans for the purchase of construction materials and the
hiring of labour. These loans proved to play an important role in the self-help
housing process, speeding up the rate of consolidation and reducing either the
need or the length of time that households had to live in temporary housing.
However, the standard of housing in the three older settlements studied was
similar, indicating that over time many households who have not taken out
loans manage to catch up to the standard of housing of those who have. Once
households constructing in illegal settlements felt secure that their houses
would not be demolished, they began to invest in permanent structures.

Households adopt many coping strategies to reduce the costs of construction
including the use of second-hand materials, gifts, direct participation in the construction process and the establishment of home-based enterprises. These
strategies again vary according to the resources available to a particular household, which are influenced by, among other things, their income, employment,
contacts and the gender of the head of household. The priorities and preferences
of households inevitably vary, and as Kellett and Garnham (1995) have argued,
cultural and social variations can also be crucial in conditioning the way in
which groups and individuals take decisions about housing.

Attempts by community-based organizations, government and NGOs to
help to produce cheaper housing solutions have experienced a range of
difficulties. Many of these stem from the fact that the self-help housing process
is a very individual one, the success of which depends on households being able
to tap all the resources available to them. Thus, when attempts are made to
formalize the self-help construction process many of these informal arrangements are prevented from operating. This conflict is exposed by, for example,
looking at the impact on women of the supposedly gender-neutral strategies

introduced by various groups and organizations. In general, female-headed households were unable to, or had difficulties in, participating in many of the projects. In short, to be successful, households, community organizations, NGOs and others, must take into account the informal nature of the finance mechanisms and construction methods of the self-help builders, and the differing circumstances of the households involved.

NOTES

1 The term 'strategy' is used here to denote the different ways in which self-help households construct their dwellings.

2 The bulk of the fieldwork was carried out in 1987 as part of an ESRC-funded PhD linked to the work of Professor Alan Gilbert. In 1997 a return visit was made to Pereira with funding from the Carlsberg Foundation to investigate the housing consolidation process.

3 In the three older settlements a 20 per cent sample was employed resulting in 30 questionnaires being conducted in El Plumón, 35 in Las Mercedes, and 40 in Leningrado. In Villa Santana a 10 per cent sample was used and 95 interviews were conducted. In 2,500 lotes, a 15 per cent sample of resident households, 50 in total, were questioned.

4 Ten years later, the legalization of El Plumón was still being discussed, but with 1997 being an election year the inhabitants were hopeful of receiving their papers within a few months.

5 In the study, the 'household' was defined as those people who lived together and pooled their resources. It is recognized that the household is not an undifferentiated socio-economic unit, is a site of both conflict and co-operation, and is constantly being transformed due to changes in the domestic cycle and wider processes of social and economic change (Kabeer 1991, Robertson 1991). As Varley (1994) has argued, there is a complex interaction between definitions of the household and the housing process as neither households nor houses are static entities. This makes it problematic to relate housing type to household type unless there is detailed information on changes in both over time. There is, therefore, no attempt to compare statistically the standard of housing and the strategies of female-headed households with those of households with both male and female heads as it cannot be assumed that a house has been built by the those members currently resident. For example, several instances were encountered where a woman had been recently left by her husband; hers was therefore classified as a female-headed household, but the husband had been very instrumental in the construction of the house.

6 It has since been replaced by Instituto Nacional de Vivienda de Interés Social y Reforma Urbana (INURBE).

7 Most households could not distinguish between the ways in which they raised finance to purchase land and construction materials and to pay for hired labour.

8 Baross and Mesa (1986) also found that the initial family investment in houses in Medellín was substantially different between invasions and pirate urbanizations. They found that the lack of security of tenure in invasions acted as a strong disincentive for households to invest in their houses for the first two to five years.

9 SENA employees did give advice of a more practical nature on site and a few dedicated employees spent their Sundays at 2,500 lotes. However, there was a limit as to how many places they could be at one time and they tended to develop an allegiance to certain communities.

10 Ironically, it was a small group of women who were the most active in the course in which I participated.

9

HOUSING FINANCE FOR REFUGEES AND FORCED MIGRANTS IN RUSSIA

Beverley L. Drought and Gareth A. Jones
University of Wales, Swansea

INTRODUCTION

Refugees and forced migrants present particular challenges to national governments and international agencies as the rapid movement of large numbers of people with limited economic resources place a strain on housing policies. Although reliable data on global refugee numbers are difficult to come by, recent estimates converge on a figure of 15–20 million refugees and, perhaps, an equal number of internally displaced persons (Cernea 1995, V. Robinson 1995, Rogers 1992, UNCHR 1995). In most developed countries, the response to refugee crises has been either to integrate the refugee cohorts into the host population or, increasingly, to pursue policies which promote settlement in countries of 'first asylum' (Escalona and Black 1995, V. Robinson 1995, Rogers 1992). In developing countries, the response has been to establish camps or 'safe areas' with either international or NGO assistance, or to turn a blind eye to self-settlement (Bascom 1993, Rogge 1987). There is, however, a growing awareness that inappropriate settlement policies can induce further impoverishment for people in already difficult circumstances.

One recent example of uncontrolled population movement is the return to Russia of approximately 3 million ethnic Russians from the fourteen non-Russian republics of the 'near abroad' following the break-up of the former Soviet Union and the subsequent re-ordering of ethnic hierarchies (Aasland 1996, Zayonchkovskaya, Kocharyan and Vitkovskaya 1993). Unlike refugee flows in either developed or developing countries, the return of the ethnic Russians presented a very different problem for policy-makers.[1] First, unlike most forced migrants and refugees, those leaving the 'near abroad' were not seeking 'third country' asylum but were returning to their ethnic homeland in search of 'permanent' accommodation and employment, and with some expectation of being met by a welcoming and ordered programme of resettlement.

Second, the forced migrants and refugees arrived in Russia at a moment of economic and political transition which probably has few parallels elsewhere in the world and which had profound consequences for the provision and acquisition of housing. Specifically, Russian housing policy was undergoing a massive paradigmatic shift that witnessed 'a move from regarding the provision of shelter to citizens as one of the state's principal objectives to seeing housing as a commodity which the citizen must acquire in the market place' (Andrusz 1992: 210). As of 1995, however, traditional central planning methods were breaking down faster than market mechanisms were emerging to replace them (Renaud 1995: 1261). The forced migrants and refugees, therefore, were confronted with a situation in which housing production was in decline, a significant proportion of public housing was being privatized according to rules which disqualified many returnees from *direct* participation, and (until 1994–5) there was no private housing finance system in place (Daniell and Struyk 1997, Kosareva 1993, Renaud 1995).[2] Administering the provision of new construction, privatization and finance reform was an ill-prepared local government system facing political and fiscal crisis.

In order to assist the forced migrants and refugees with housing, the Russian government attempted to redirect existing housing policy to cope with the extra pressure, but with only moderate success. Taking a lead from the migrants themselves, however, the government also introduced an innovative method of self-settlement known as 'compact settlement'. Although termed 'settlements' these are in fact micro-enterprises consisting of a group of forced migrants, most of whom were work colleagues in the former Soviet republics (the 'near abroad'), and who have organized their own housing and economic resettlement with financial assistance from the government. As such, the compact settlements had the potential to operate along many of the principles commonly identified as critical to successful migrant and refugee self-settlement: an efficient targeting of resources, minimal disruption to refugee lifestyles, raised levels of personal control and lower institutional dependency (Bascom 1993, UNHCR 1995). Furthermore, they make a direct link between micro-enterprise development and the provision of housing similar to that attempted by many NGOs working in micro-finance (Jones and Mitlin this volume).

The assistance provided to the compact settlements also appeared to uphold many of the policy characteristics associated with effective housing finance programmes. The programme made loans to target individual migrant households who possessed at least one income earner, provided relatively transparent subsidies and encouraged matching finance from the household. Nevertheless, we identify a number of shortcomings with the design of the finance programme and the methods by which the settlements have been administered, which have resulted in them failing to provide housing. In particular, the best intentions have been undone by political interests which the programme failed to take into account. We argue that housing finance for refugees and low-income households more generally needs to acknowledge how the 'aims, methods, personal development, rivalries within and between different organizations' affects the outcome of resettlement policies (Escalona and Black 1995: 383).

Fieldwork for this chapter examined the resettlement of ethnic Russians from the republic of Tadzhikistan which in 1989 (date of the last Soviet census) had 399,000 ethnic Russians, but which according to the Tadzhik embassy in Moscow in 1996 possessed no more than 70,000. Fieldwork was conducted in five compact settlements during 1995 and 1996 and consisted of ninety-four semi-structured interviews with migrants, settlement leaders, representatives from government and the NGO sector, and participant observation of settlement meetings.

HOUSING REFUGEES AND FORCED MIGRANTS: THE OFFICIAL SOLUTION

A significant proportion of the 3 million migrants to enter Russia since 1990 did so in the earliest stages of the Soviet Union's break-up and went unrecorded by the government (Zayonchkovskaya, Kocharyan and Vitkovskaya 1993). Anecdotal evidence suggests that these early returnees were among the better-off, predominantly Russian-born, population from the republics and therefore possessed a range of housing options. It is likely that many were able to transfer employment and housing through the Soviet system of internal exchange, possibly with help from repatriation societies such as *Migratsiya* in Tadzhikistan, or use graft and corruption to jump long housing queues (Morton 1987). It is also possible that many were able to return to housing which was retained while they were working in the 'near abroad', were able to live with relatives or to acquire 'second homes' (*dachas*), which witnessed considerable new investment and rising prices from 1990 onwards.

Official figures on the number of returning ethnic Russians only exist from 1992 when the Federal Migration Service (FMS) was created to assist repatriation. According to the FMS, 947,401 ethnic Russians are officially registered as having returned to Russia between 1992 and 1996. By contrast to the early arrivals, this group was subject to regulation and monitoring by the government and, from 1992, a legal division of status into either 'forced migrants', those with full Russian citizenship, or 'refugees', ethnic Russians without full citizenship rights. Although the government is obliged to provide both groups with resettlement programmes, the division has allowed it to ration assistance generally, and to provide less help to refugees more especially. The government's involvement in what has become almost an unofficial international competition to confuse and complicate refugee rights, has had important implications for the way households locate and construct housing, and how this housing is financed.[3]

These later returnees have had to overcome a number of obstacles in order to access housing. First, from 1992, four-digit inflation has eroded savings which might have been used to invest in housing. Second, the chance to sell housing in Tadzhikistan and some of the other republics was closed off due to the civil wars and as the market was flooded with a mass of Russian-owned properties. In 1992–3 apartments in Tadzhikistan were selling for less than US$ 500 whereas prices in rural Russia were US$ 4,000–5,000 for housing without electricity and about US$ 14,000 for a serviced unit. Renaud (1995) reports a price/income ratio for apartments in Moscow of 29 (compared to an average of 4–5 in developed countries), making finance an urgent necessity.

To meet its obligations to registered forced migrants and refugees the Russian government established in 1992 the Federal Migration Programme which set out a package of targeted loans, subsidies and grants as well as four housing alternatives: government-funded emergency housing; FMS-designated housing; acquisition of privatized stock; and subsidized self-settlement. In the face of political uncertainty and economic instability, however, it is not surprising to discover that housing provision through the Federal Migration Programme has been limited. From the outset, inadequate resources forced the FMS to restrict access to emergency housing to the victims of ethnic conflict and to 'socially unprotected' households, defined as single mothers, war pensioners, the disabled and households with more than three children.[4] By 1996, however, there were only eighty-nine emergency centres with a capacity to accommodate approximately 95,000 people, two-thirds of whom were from the Chechen republic.

Access to the second option of FMS housing is made available to forced migrants through the Temporary Housing Reserve and to refugees through the Special Housing Bank. In both cases, the expectation was that joint ventures between local government (*oblast*) and the local migration services would complete the construction of stalled public housing projects or encourage the sale of public buildings to the FMS. This policy, however, has had to compete with other programmes to encourage private construction firms to complete former government stock and to sell the units on the private market (Daniell and Struyk 1997). Consequently, by the end of 1995, only 7,323 units (mostly flats) had been constructed or sold to the FMS. Again, priority for these units goes to socially unprotected households, and thereafter to all forced migrants and finally to registered refugees. As of June 1996, the priority waiting list contained 12,679 households and there were a further 44,776 households on the non-priority list: that is, the lists include eight times more households than the number of housing units built or acquired in the previous four years.

The third housing option, acquisition through the privatization of local government stock, has also been problematic. Officially, acquisition of a privatized housing unit is restricted to Russian citizens, thereby excluding refugees, but even forced migrants have faced serious impediments as priority is given to those households that have been on local government waiting lists the longest:

official estimates put this figure at 10 million and many have been registered for upwards of thity years (Kosareva 1993, Morton 1987). Forced migrants who have only recently registered with the local government, therefore, are placed at a disadvantage compared with the indigenous population.

Given the failings of these housing options, most migrants have had little alternative but to live with relatives (which most do) in already overcrowded living conditions or to self-settle. Indeed, it would appear from the outset that the Russian government was aware of the difficulties of its own migration programme. Resolution 327 of the Russian parliament (1992) recognizes the impossibility of the government providing housing to all forced migrants and refugees, and suggests that resettlement be adopted using the migrants' and refugees' labour and resources. As of mid-1996, official figures record over fifty 'compact settlements', although there are probably an equal number of settlements not registered with the FMS. Self-settlement appeared to offer forced migrants the opportunity to acquire better housing than that available through the public sector, as well as employment and access to a discrete package of financial assistance. It is to the performance of the self-settlement option that we now turn.

SUBSIDIZED SELF-SETTLEMENT: COMPACT SETTLEMENTS

The Federal Migration Programme aims to encourage the direct participation of forced migrants and refugees in housing construction through a form of self-settlement known as the 'compact settlement'. Compact settlements serve two functions: to establish some form of production enterprise (examples include sewing shops, farms, aluminium purification, cable production) and to invest accrued profits into housing construction. In a few cases, the economic base to the settlement is house construction, with the profits from sales cross-subsidizing housing provision for settlement members.

From 1992, the establishment of compact settlements has been regarded as an important mechanism to solve the forced migrant housing and employment problem. Through the FMS, forced migrants are eligible to receive a package of financial assistance: a one-off grant of US$ 43 (equivalent to a single monthly minimum wage) and the option to apply for a loan of US$ 1,800 per household or US$ 2,700 in the case of socially unprotected migrants. In many cases these loans are 'pooled' by migrants and handed to a settlement leader or managed through the enterprise. Land for the settlements is usually donated by local governments, a state farm or enterprise (in return for labour or a share of settlement profits) or the pooling of the migrant entitlement of either 600 square metres of land in urban areas or 1,500 square metres in rural areas.

Once a settlement has been established, the FMS is supposed to make further loans of up to US$ 480 per plot for the installation of basic infrastructure, although the eligibility criteria for the loans appear to be vague and unknown to many forced migrants. Finally, there is a suite of loans available from the FMS and the Federal Employment Service, and an interest-free 'revolving fund' organized by the United Nations High Commissioner for Refugees (UNHCR) and a Russian NGO known as the Compatriots Fund, for the creation of additional workplaces in the micro-enterprises. On paper, therefore, compact settlements appear to offer forced migrants substantial additional resources to assist in the acquisition or construction of housing.

COMPACT SETTLEMENTS: AN ADEQUATE SOLUTION?

During 1996, the experience of four compact settlements was looked at in detail. Two settlements were selected because the central enterprise was based upon housing construction. To join these settlements migrants were required to invest in and work for the construction company, in return for which they would receive a subsidized house. One settlement, Xoko, employs 950 forced migrants and is the largest compact settlement in Russia; the second, Novosel, was set up to build housing for over 1,000 migrants. Both Xoko and Novosel have received funds from migrant members and FMS loans which have been pooled for collective use. By contrast, the other two settlements had manufacturing as their economic base: Gulshan, which began with sixty-six former employees of the Chemical Institute in the Academy of Science in Dushanbe (capital of Tadzhikistan) and has diversified from the purification of aluminium into the production of air and water filters; and Voronezh Cable, which was established when the former director of Tadzhik Cable moved to Russia and brought with him fifty-six employees. Only some of the migrants in Gulshan and Voronezh Cable have received FMS loans and these are administered individually.

In all four compact settlements it is clear that despite a range of financial assistance, housing construction has been very limited. Indeed, from the figures it is impossible to tell which settlements have housing as their principal activity and which do not. Of the four settlements, Xoko is considered by the FMS, the UNHCR and the International Organization of Migration (IOM) to be the most successful compact settlement in Russia. Albeit with the assistance of unparalleled funding from federal and NGO sources, Xoko has been one of the few settlements to provide employment and temporary accommodation to its members, as well as constructing an additional forty-five flats and twenty-nine cottages. By contrast, Novosel has received far less funding and although housing construction began in 1992, along with a carpentry and a sewing workshop, building stopped when further funding was cancelled in 1994 amid accusations of embezzlement. By late 1996, the settlement had built only five houses with an additional fifteen individually built units under construction. Of the original 120 households to arrive in Novosel from Tadzhikistan only sixty remained. In Gulshan, two houses have been built and Voronezh Cable has purchased eleven flats. In the meantime, in Xoko and Novosel the migrants live in *vagonchiks* and *botchkas* (large metal crates or barrels) donated by the FMS, and members of Gulshan and Voronezh Cable live in rented accommodation or *obshezhitiya* (communal flats). Below, we outline some of the more generic reasons why these compact settlements have failed to convert the loans, subsidies and grants into adequate housing.

The lack of an economic base and external dependency

A notable feature of all four compact settlements is the absence of a core enterprise which is able to compete in a Russian economy undergoing restructuring and recession. Gulshan, which is based on a successful high-grade aluminium purification company, had problems when demand declined as the military budget dried up in 1993. In Novosel, the absence of an economic base is even more obvious. The settlement has no external market for the clothing, foodstuffs and construction materials being produced, so that, since 1993, the housing loans have been subsidizing the micro-enterprises. Faced with a declining economic base, some settlements such as Voronezh Cable have diversified, in this case into the sale of cement and fertilizer, as well as the production of different types of cable. In the absence of surplus capital, however, diversification has been achieved by diverting funds away from their intended use. As the leader of Gulshan put it:

> We knew when we accepted the loan that it would be impossible to keep their terms, but we had no choice. We needed the money to buy new equipment for the production of air and water filters.

The dramatic failure to create a viable economic base has nurtured a dependency on external finance. Rather than the sustainability which was expected after the start-up finance had been delivered, settlements have become increasingly reliant on external sources and we doubt if any of the settlements investigated would break even according to standard accounting criteria.

This sense of financial dependency on external sources for employment is carried over to migrant attitudes toward housing provision. Many migrants insisted that the government should provide permanent accommodation and questioned the self-settlement option. Indeed, settlements seemed to be divided between migrants (notably those involved in house construction) who felt motivated to overcome the economic difficulties and those who, in the words of the first group, preferred to 'sit and complain'. In expressing these views there was a clear age component with the older migrants displaying higher levels of confidence and adaptability than many of the younger ones. As one male member of Novosel put it:

There is a need to change the psychology of migrants. They need to accept that they have moved from somewhere which has asphalt roads to dirt tracks. They need to work at their new lives. Old people have changed quickly. I hope the rest will follow. It is the elders who show the youngsters what to get from the land. These are the people with life experience. They all survived World War Two, so now they can survive this. . . . People need to understand that they have to accomplish things for themselves and not sit and wait; that no one will bring it to them.

Another member of the same settlement explained:

I have built my own house, but the youngsters here still have their Soviet mentality. They can only wait and hope to receive flats. There are people here younger and stronger than me but all they do is sit in the *vagonchiks* and wait and do nothing. We have to change their mentality. . . . You need to work for yourself.

In their defence, other households argued that their minimal income made it impossible to invest in housing and that any surplus was better spent on food. Others argued that it is difficult enough to find the time and energy to grow their own food, without having to construct their own housing. Such sentiments are partly fuelled by an internal system of migrant dependency which the organization of the compact settlements and the FMS have fostered.

Leadership and internal dependency

As the weaknesses of the economic base and the inability of compact settlements to deliver sufficient quantities of housing have become evident, migrants have begun to point the finger at the role of leaders. In most settlements the completed housing was to have been allocated to migrants according to length of residence and quality of work. In reality, what little housing exists is allocated to migrants whose skills and expertise are most required by the enterprise. This has left one woman to complain:

We are pensioners. Because of this we are of no use to Bullin [the leader]. He needs to give the housing to workers who are of value, whose expertise is of value. He can hire anyone to do my job so I don't qualify for housing. . . . I don't know how Bullin distributes the flats but there are a lot of 'dead souls' on the list. It certainly is not dependent on the amount of time you have been working for Bullin, otherwise I would have received one of these flats.

This same respondent described the compact settlement experience as a 'type of slavery', a sentiment that had parallels in other interviews. In all settlements, migrants claimed that the leaders were using the promise of housing as a bribe to guarantee high-quality labour in the enterprises at minimum cost. They argued that instead of providing housing and welfare the compact settlements promote a work ethic in return for wages of US$ 40–80 per month. Some migrants likened working in the compact settlements with the promise of housing as being on permanent trial, a feeling which was felt to be psychologically damaging and exhausting. One woman in Voronezh Cable expressed the view that:

> Bullin [the leader] knows we are prepared to work hard for him and under the most difficult of conditions because our one goal is to receive housing. He can then make us work as hard as he likes. We have no choice but to remain with him. We won't receive housing from anywhere else.

And another in Xoko that:

> People are tired; tired of waiting; tired of poor living conditions; tired of being scared that they will be thrown out.

As many migrants had 'pooled' their FMS loan with the leader, they had little opportunity to leave.

The pooling of loans as a means to achieve a critical capital base for the settlements was also resented by migrants when it became obvious that the enterprises were failing to make sufficient surplus to fund house construction. In 1996, there was a growing argument in favour of the collective approach giving way to individual house construction. The following view from a woman in Novosel was common:

> If you had finance to construct your own housing you would use it wisely. If you had it in your own hands. People spending money on your behalf are likely to do so carelessly. You would manage your own resources better if you are given control. Nothing is ever done correctly if you are being controlled by others because they don't know your needs.

There is also some evidence that leaders have used intimidation against the migrants, using housing as a lever to assure compliance. As a male migrant in Xoko put it:

> People are afraid here. If they complain then they would be refused a job, a *vagonchik* or a house. If we ever opposed the leaders we would be asked to leave the organization. . . . The sense of dependency on the

leaders is unbearable for us. We feel as if we are not free. We are placed into a situation of dependency. The leaders of Xoko exploit our situation. They did their utmost to create these conditions of dependency so that they can benefit.

In Xoko, Gulshan and Voronezh Cable, leaders have warned that at least one member of every household must continue to work for the enterprise or else the household forfeits the right to a house or their room in the communal flats. More importantly, if a migrant leaves, they are told that they will lose their registration and thus void their right to education, medical care and social benefits from the local government. It is a heavy irony that migrants keep silent, even in the face of alleged wrong-doing, for fear of losing a set of potential rights or housing which has not been built. This is most obviously the case in Novosel where rumours of leader embezzlement were rife. As one woman explained:

> When we worked in Novosel we could not speak out against Galina [the leader] because she was the one in control of resources. She could have fired us and then we would have had no chance of receiving housing. We knew things were wrong but we all kept silent.

Against her powerful position as leader of Novosel, however, Galina has had to counter claims that she is too weak for the post. According to some of the men, Galina acquired her position because of her position in the Communist Party and a state deputy in Tadzhikistan under a political system which gave considerable emphasis to gender equality. But now, the men argued, she was not tough enough for the 'new' Russia where authority is based upon ability and not party. During interviews, the men pointed to her resorting to 'female traits' such as crying in public, her refusal to pay bribes to officials to get her way and the personality clashes with the heads of the local FMS office and local government, both women, as indications of her weak leadership. While Galina's leadership qualities and her gender are not sufficient to explain every problem encountered by Novosel, it is our impression that the general lack of accountability among settlement leaders is a worrying feature of the Federal Migration Programme.

Gender bias in housing finance disbursement

The actual method of making loans to the migrants has implicitly discriminated against certain household structures. While the FMS avoids adopting a limited definition of a 'household' as nuclear, in practice, the loan-making procedure discriminates against female-headed households, which represent about 8 per cent of forced migrant households, and divorced, separated or widowed women who are forced to live with relatives in extended arrangements.

145

Although not disbarred, female-headed households have little incentive to apply for loans which are insufficient to hire construction labour and require that the household contribute additional resources in order to purchase materials and install services. Of course, the female-headed households are less likely to possess the extra resources to start construction due to the absence of a second income earner. Moreover, the women that do work earn lower incomes and have less stable employment than their male counterparts because the settlement leaders prefer men for the better jobs. In a few cases, the precarious economic base of the compact settlements has forced leaders to prevent female-headed households from joining, arguing that they have less to contribute. The possible exception is Novosel, where the leader, a woman, appears to have a more positive attitude to accepting female-headed households, but has been attacked by migrants for treating the settlement as a charity.

The gendered nature of the FMS loans is also a consequence of the financial structure of the agency. This would appear to take two forms. First, the FMS has a rule of granting one loan per household which means that it is difficult for women residing in extended arrangements or whose partner/husband resides elsewhere due to work commitments, from acquiring a loan in order to establish an independent household. The one-loan rule also prevents members of households that have received a loan from leaving as they are ineligible for a second loan. Thus, households which set up extended arrangements on arrival from the 'near abroad' find it difficult to finance new household formation.

The second form which discrimination takes relates to the politics of the FMS itself. Under pressure to assist as many migrants as possible from within a limited budget, some sources suggested that the FMS has given preference to the larger and extended households. As most female-headed households would qualify for a US$ 2,700 loan as a 'socially unprotected' forced migrant, they would also cost the FMS more to serve. Although, the FMS's data are not organized in such a way as to cross-reference whether female-headed households are proportionately less likely to receive a loan, the general picture is that subsidized FMS loans are benefiting those migrants who can already afford to construct or purchase housing.

Inconsistent housing finance policy

The problem of FMS loans being insufficient to construct the type of housing to which the migrants aspire applies to all households, but its effect has varied over time. For the earliest recipients of loans (in about 1993–4) the problem was that three-digit inflation reduced the real value of the loan as construction costs rose. Thus, while the rampant inflation meant that the unadjusted interest rate became steeply negative and many of the ten-year loans were repaid in just a few months, there was little housing to show for it. For later recipients, the value of the non-indexed loan was further reduced while falling inflation reduced the negative points spread and created repayment difficulties. In 1995, an FMS loan was barely sufficient to cover the cost of house foundations and therefore required households to commit extra resources. As one man in Novosel remarked with bitterness:

> Why should I have a loan only to build house foundations? What am I supposed to do, put a tent over it and pay off my mortgage being thankful that I have this? No thanks.

According to the migrants, the rigid structure of FMS loans has actually worked against the construction of housing. One problem is that FMS loans are not distributed as a cash advance, but rather as reimbursement from a bank account against designated building costs on receipt of a contract or invoices. Not only does this take time (with the consequent effect on the loan value with inflation) but it hampers those migrants who would prefer to finance activities other than housing in the short term: notably, to fund a child's education, to buy food or to insulate a *vagonchik*. In order to use the loan in these ways, migrants have had to persuade construction materials suppliers to forge receipts (in exchange for vodka or small bribes) which could then be presented to the bank.

As incomes have fallen or been unstable, and the real value of the loans has declined, in order to complete a house migrants have had to find additional funds. In reality, this means that FMS loans which are provided for the purposes of housing are 'adapted' to generate income. One of the most successful cases involved a woman in Novosel who purchased a car to transport meat from local farms to Moscow, and to deliver bread and meat to nearby villages. The resulting profits allowed her to pay back the loan, pay for her daughter's college education, to insulate the *vagonchik* and conduct initial work on a house. Although denied permission by the local government, the woman now has plans to establish a bakery in Novosel.

Even with such 'adaptation', however, the evidence is that the FMS finance programme has been translated into housing is, at best, limited. Therefore, while in 1995 the FMS claimed to have distributed US$ 24.9 million as individual loans, a sum that would be sufficient to provide one loan to approximately 12,900 households, these figures are considerably more than the actual number of houses built or improved.

The politics of envy

Other migrants refused to blame the settlement leaders for the lack of housing and pointed instead to the resistance from local government and the resentment of officials to incomers: a widespread feature of Russia's political structure that pre-dates the formation of compact settlements (Andrusz 1992). These officials gain influence from the inconsistent structure of the Federal

Migration Programme and its incompatibility with the general policy to promote greater decentralization. One example is that while forced migrants are
entitled to FMS loans and other benefits, legislation enacted in 1996 entitles
all Russian citizens to a local government housing subsidy of between 5 and 70
per cent of total costs.[5] As a result many migrants are not applying for the FMS
loan or have decided to delay self-financed construction in case this disquali-
fies them from the more favourable local government subsidy. The availability
of a parallel system of housing finance, therefore, has stalled housing construction even though, to date, no migrant in the four compact settlements
studied has received a local government subsidy.[6]

The apparent inconsistency of the federal and local housing finance system
is related to a growing trend for local governments to prevent forced migrants
from being eligible for public housing and services. Methods include the
imposition of punitively high fees charged to migrants to register with local
governments, without which households can be denied access to health and
education services. At issue here is the method by which compact settlements
receive funding. As federal projects, most funding comes to the settlements
without passing through local government. Such 'goal-oriented' funding
deprives local officials of opportunities to embezzle funds so that, at every subsequent stage, they attempt to exert influence over the settlement process by
withholding land allocations or grant migrant registration. One response by
settlement leaders has been to build extra housing units away from the compact settlements for acquisition by local officials.

The model 'goal-oriented' funding and compact settlements are also perceived by officials as a threat to existing political networks. In particular, as
many of the settlement leaders were leading party apparatchiks or KGB officers in the 'near abroad', local officials and politicians fear that the settlements
will serve as a power base from which migrants will attempt to exert influence.
Local officials, therefore, have had a political interest in assisting the failure of
the settlements whenever possible.

A final factor in this equation is the competition felt between locals and
migrants over claims to a Russian identity. Local officials regard the ethnic
Russians from the 'near abroad' as less than Russian, a feeling confirmed by the
'forced migrant' label which implies a distinction from full citizens, and they
enjoy the opportunity to get one over on them. This envy is matched by the
migrants' low view of the locals, who they regard as having fallen into corruption, alcoholism and immorality (Aasland 1996). The migrants have
disdain for their current living conditions which is far removed from what they
experienced in Dushanbe and do not understand why officials should object to
their attempts to improve things. As one man in Gulshan put it:

> The main problem we have with the local population is their resent-
> ment and their envy. They see the type of housing we want for ourselves
> and don't understand why. They cannot see why we do not want to live

in the same conditions they do – in a single room with no conveniences. We are not satisfied with this type of housing and why should we be satisfied? We want a private house, with several rooms, a toilet, running water, a sewerage system. We consider these conveniences as normal, and not as extravagant . . . we respect ourselves and our families. Why should we live in such appalling conditions especially when we have the chance to build better and improve our position?

This envy and sense of competition between local officials and the migrants leads to delays, increased costs and uncertainty in the self-settlement process.

FROM SELF-SETTLEMENT TO SELF-FINANCE

With the near collapse of the economic base of a number of the compact settlements and the general inability to deliver housing, many migrants have attempted to find alternative sources of income and to construct permanent housing on an individual basis. Migrants living in *vagonchiks* or rented accommodation have used their building plots to establish vegetable gardens and introduce livestock. Such work is attractive to nuclear and extended households, with both men and women sharing the workload. For those households with the lowest incomes or female-headed households, however, this method of raising income is difficult, especially as only a few applied to receive a free plot because of upkeep costs.

Others have raised income levels by finding temporary or extra employment. Building work is a logical source of employment for the younger men and one which brings its own set of opportunities. One group of men from Novosel secured contracts in 1996 worth US$ 6,000 to renovate and construct dachas in Kaluga. One member has bought a small lorry to transport sand from a local quarry, and charges Muscovites US$ 80 and migrants US$ 40 for a delivery of sand when the actual cost is a bottle of vodka to bribe the guard at the quarry. These extra earnings have allowed the man to rent a concrete mixer and he now produces bricks. Other jobs include an electrician in Xoko who mends the televisions and radios of local residents, while one woman supplements her income by making clothes for contacts she has in Moscow. In three weeks she can earn as much as US$ 270, which has been used to buy food and clothes. One household made US$ 430 in 1995 by collecting mushrooms in the forest and selling them to a firm for export to Germany and another migrant attempted to grow marijuana on his family plot in Tadzhikistan.

More conventionally, male heads of household and young males in extended households have been forced to accept seasonal work away from the compact settlements, raising incomes but reducing the time available for construction. Working away from the area is the preferred option for some of the younger males so as to reduce the amount of money spent on daily travel, and to maximize the time and disposable income to invest in housing construction. As a man in Novosel put it:

> If I am working, I need to know what I am working towards. Working in a factory means no hope. It is not an answer. It means you are living from hand to mouth and nothing else. . . . If I had permanent employment in Kaluga I would not have the time or the money to help my parents construct housing here in Novosel. . . . We could all find employment and rented accommodation in Kaluga, but this is no way out. We would not have any money or time to construct our own housing.

Instead, this migrant secured regular work (for three months) in Smolensk which allows him to return to Novosel for short periods with enough money to undertake construction. Working away does, however, have its penalties, in particular the payment of 'roof' money for local mafia protection. In addition, economic and political conditions have made it difficult to secure employment outside the immediate area of a compact settlement. In 1995, for example, one source of male employment, working as casual labourers renovating houses in Moscow and constructing dachas, dried up as owners awaited the outcome of the elections.

As the economic conditions in the settlements have got steadily worse over the past four years, the departure of more men to find work has created 'temporary' female-headed households. Working away from the compact settlement has not been an option for women, who have responsibility for domestic jobs, in particular the collection of water and fuel to heat the *vagonchiks* and *botchkas*. Although women do work for the compact settlement enterprise, sometimes covering for their partners when they are away, and a number have professional jobs in local towns, they have fewer opportunities to acquire more lucrative employment and the extra jobs tend to confirm gender stereotypes (food preparation, clothes making).[7] Economic coping strategies caused by economic recession, the poor economic base of the compact settlements and the inappropriate financial support of the FMS, have placed additional burdens on the household.

One group, however, has emerged as well placed to finance self-help housing. Able-bodied, predominantly male, pensioners have some independence from younger household members through possession of a state pension which is distributed efficiently compared to the wages of private- or public-sector companies. Moreover, pensioners can be among the highest income earners in the settlements: for a married couple a pension can be as much as US$ 163 per month; socially unprotected pensioners (invalids, the disabled and participants of the Second World War) qualify for US$ 120 and military pensions can be as high as US$ 160. It is the pensioners, therefore, who are emerging as the self-builders in the settlements. A situation has developed in which having a relative of pensionable age within the household is considered to be an asset.

Forced migrants and NGOs

Until 1996, international and national NGOs played a limited role in the self-settlement of forced migrants and refugees in Russia. While the IOM provides grants and the Compatriots Fund and UNHCR have supported resettlement through the provision of start-up capital loans to migrant enterprises, allocations have been sporadic, and the subject of alleged embezzlement by NGO officers, and project performance has gone unmonitored.[8] Nevertheless, despite this dubious track record, a recent conference organized by the IOM and the UNHCR agreed a programme of action that contains a series of measures to solve the employment problems of forced migrants through micro-enterprises and better use of revolving fund finance (UNHCR 1996).

Should the programme operate by providing enhanced support for indigenous NGOs set up by and under the control of the migrants, there is some cause for optimism. An increasing number of migrant-run NGOs in Russia have been registered in order to co-ordinate the activities of migrant-owned enterprises and to develop micro-finance schemes. One example is Doveriya in Nizhni Novgorod, which has received US$ 70,000 from Opportunity International (an American volunteer organization) and the World Bank, and which distributed 226 micro-finance loans (including ten to forced migrants) of US$ 300–900 between 1995 and 1996. A second example is the Association of Saratov Spring, which has already campaigned successfully to allow migrant enterprises to tender in federal competitions, and in 1995 distributed a total of US$ 860,000 between enterprises (both migrant and indigenous) for the creation of 1,155 workplaces. The association has plans to establish a revolving fund (*Solidarnost*) with financial support from the UNHCR to provide employment and housing for refugees and forced migrants. Two companies, a clothing firm and a manufacturer of television sets, have guaranteed *Solidarnost* a three-month investment programme to create additional workplaces. Two groups led by female forced migrants from Azerbaijan and Tadzhikistan, one of which was set up from a failed compact settlement, have already approached the association with plans to establish housing co-operatives. These types of initiatives promise to make direct links between micro-enterprise and housing with less reliance upon the FMS loans, be subject to greater internal monitoring and external scrutiny, and be less gender-blind in their operations.

CONCLUSION

During the early 1990s the Russian government was faced with the enormous problem of how to resettle approximately 3 million ethnic Russians from the 'near abroad'. Although the government had expected conventional policies to cope with a large part of the housing problem, through the FMS it nevertheless appeared to be adopting some innovative strategies. These strategies included the provision of lines of finance for housing and employment, and the promotion of compact settlements. What the government could not have known was the extent to which these strategies would be affected by the transition to a more market-oriented economy, nor the contradictions that this set in train. The subsequent failure of the FMS migration programme has left many of the poorest migrants without housing, jobs or adequate welfare provision. As a woman in Novosel explained:

> We all originally thought that compact settlements were the answer. They allowed Russians from the republics . . . to buy housing together as a group; to come together and build their own form of settlement and be employed at the same time. Employment would come initially from the housing construction and then later from the enterprises and services the settlement would need. People would also be able to find employment in local markets. But this has not been allowed to happen. We have been strangled [by the FMS].

Almost all migrants and refugees are worse off today than they were prior to leaving Tadzhikistan.

The experience of housing finance, particularly through the compact settlements, indicates the extent to which programmes need to match local political conditions. In Russia, leader exploitation of migrants as a source of cheap labour appears to be widespread and leaders have regarded the housing loans as a 'soft' resource that can be used with little accountability. In addition, FMS policies have run into the intransigence and vested interests of local government officials who have been reluctant to provide land for the settlements, to register the members or, since 1996, to make the housing subsidy available. Future finance programmes should learn from these mistakes and seek to lock-in those groups able to dictate success or failure, possibly through the organization of partnership or trust arrangements.

The experience of the compact settlements also provides insight into the relationship between community and housing finance. Like many communities in developing countries, the groups that formed the compact settlements were created out of adversity and, despite a common ethnic and migrant identity, it has proven difficult to maintain solidarity as conditions deteriorate. In particular, problems of finance allocation and the transparency of decision-making were sufficient to undo much of the sense of unity. The low priority afforded by the FMS to maintaining and monitoring the settlements, and the general absence of NGO involvement which might have assisted with community-building initiatives, meant that by 1996 some of the compact settlements were close to collapse.

Finally, by failing to target finance the government has not managed to assist those most in need such as refugees and female-headed households, but has favoured those forced migrants who were already in the best position to adapt to a return to Russia. All migrants in the compact settlements, however, have suffered from the catch-22 that wages are kept low in order to provide a surplus for house construction, which means that households are unable to contribute sufficient personal funds to self-build once it becomes clear that the promised housing will not be available. Consequently, only able-bodied migrants with additional income-generating opportunities or those with pensions can afford to construct adequate housing.

NOTES

1 There are, perhaps, some broad similarities with the migration between India and Pakistan during Partition and the integration of ethnic Germans from eastern Europe into a united Germany.

2 It is curious that the many studies of housing privatization in Russia have made little or no mention of the additional 0.5 million households entering the country from the 'near abroad'. Data show a significant increase in privatization rates during 1992, especially by pensioners, which may have been partly caused by returnees convincing their parents to acquire apartments.

3 Although it is almost fifty years since the adoption in 1951 of the UN Convention on Refugees, many governments, multilateral agencies and academics adopt different and inconsistent definitions of 'refugee', 'internally displaced person' and 'forced migrant' (Escalona and Black 1995, Rogers 1992).

4 According to the Federal Migration Service (FMS), about 20 per cent of forced migrants and refugees are classified as socially unprotected. Although emergency accommodation is supposedly for three months, many have been resident in the centres for much longer.

5 In some cases, this loan can be topped up with local government mortgages. The Kaluga government, for example, is co-operating with federal banks and the Ministry of Finance to introduce mortgage schemes for residents in possession of a government subsidy or for individuals who already own property and are prepared to use it as collateral.

6 In the conspiratorial environment of the compact settlements and given their experience of local officials, some migrants argue that the promise of housing subsidies is a deal between the FMS and local government to reduce the number of migrants applying for FMS loans.

7 Migrants are under no illusion that the men are privileged in working away from the settlement or household, living in single-room dormitory accommodation in Moscow for many weeks at a time.

8 According to an International Organization of Migration (IOM) Newsletter in 1994, the Compatriots Fund claims to have assisted over 700 micro-enterprises and 250,000 refugees and migrants. Yet, in our work in five Compact Settlements, NGOs and migrant societies, it seems clear that these claims are generous and difficult to substantiate.

Part III

GENDER, NGOs AND MICRO-FINANCE

10

INTERPRETING GENDER AND HOUSING FINANCE IN COMMUNITY PRACTICE

The SPARC, Mahila Milan and NSDF experience

Sheela Patel
The Society for the Promotion of Area Resource Centres, India

INTRODUCTION

Housing finance is one piece of a complex jigsaw puzzle called 'habitat'. It is a puzzle which presents a major challenge not only to low-income groups, but also to city officials, public and private finance institutions and NGOs. There is an ongoing debate in India about who has the right to assemble this puzzle; is it to be completed together by all parties or are the pieces to be fought over? This chapter, written on behalf of the Society for the Promotion of Area Resource Centres (SPARC), Mahila Milan and the National Slum Dwellers Federation (NSDF), seeks to illustrate how low-income groups, especially women, form a nucleus and are a critical factor in the struggle to resolve the housing finance puzzle.

Given the dearth of formal housing finance and the cost of informal finance, the Federation has encouraged low-income communities, and especially women, to initiate savings and credit programmes which will meet their requirements as well as create more extensive and decentralized methods of loan delivery. Although the Federation's experience in housing finance is still limited, over the last decade it has succeeded in putting into place various internal and external arrangements which have enabled large numbers of urban low-income communities to seek secure shelter.

This chapter begins by briefly examining the housing situation of low-income households in Mumbai (formerly Bombay) before describing the alliance between SPARC, Mahila Milan and NSDF and analysing the benefits of the Federation model. The chapter then examines the type of finance activities undertaken by the Federation and looks at how low-income households sustain their involvement in what is often a very slow process. It concludes by emphasizing the key role that the Federation plays in mediating between low-income groups and the government.

LOW-INCOME HOUSEHOLDS AND THE HOUSING PROBLEM IN MUMBAI

With 75 per cent of the national population residing in rural areas, resolving the urban housing problem has rarely been top of the political agenda in India. Yet, despite the low priority given to urban problems, 225 million people in India live in urban areas and of these some 20–50 per cent live in informal settlements (Patel and Burra 1994). Nowhere in India is this policy contradiction more evident than in Mumbai, the country's financial centre, a major port and the capital of the most urbanized state in the country, Maharastra. Here, 5 million of the city's 10 million people live in slum tenements and squatter settlements (Patel and Burra 1994).

Official attitudes to the presence of low-income men and women in urban areas have varied over the years. Initially, the state government of Maharastra saw itself as a key player in the supply of housing through the Maharastra Housing and Area Development Authority which received funds for the production of both housing and infrastructure.[1] However, the authority's track record has been dismal: it produces no more then 5 per cent of Mumbai's housing requirements, at high costs in real terms, and allocates much of the housing stock to better-off households. Consequently, not only has the government been unable to deliver sufficient and affordable housing, it has also absorbed all resources allocated to house low-income groups.

Those unable to receive public housing have had to rely on sporadic and often inconsistent government policies which have ranged from slum clearance to their recognition and improvement. In Mumbai, older more consolidated slums no longer face the prospect of demolition because they represent critical 'vote banks' for politicians who seek to 'protect' them as the government needs to demonstrate 'due process'. As a result, some amenities and services have trickled into these areas even though such improvements reach very few households and the quality of basic amenities is sub-standard (SPARC 1996). Newer slums have also been granted government recognition once they have earned an 'endurance badge' (currently those established before 1 January 1995 are acceptable to the Mumbai authorities), an approach that deprives settlements of the legitimacy necessary to upgrade housing.

The issue of legitimacy is an important one as there is increasing recognition of the fact that a secure urban address provides a basis for asset formation for low-income households in Mumbai. At present, although low-income groups provide 80 per cent of the national housing stock, the constraints under which they make this investment do not allow for consolidation beyond a certain point. In the case of pavement-dwellers, for example, the constant threat of impending demolition means that their tent-like structures are ready to be dismantled at a moment's notice in order to hide the material and avoid confiscation. Such actions, and the constant maintenance required due to the monsoon season, means that over a twenty-year period, the very poorest of households spend the equivalent of the deposit on a loan and interest repayments sufficient to buy a 60 square metre house. Low-income households are, therefore, building expensive but low-quality housing. Moreover, they are not constructing in a manner which enables them to accumulate a valuable asset. Each household makes small improvements when they can afford to and do so by recourse to informal finance, often provided at very high interest rates. Consequently, while each household invests considerable sums in housing construction, over time they get less for their money.

The foregoing discussion highlights the need for the government to re-examine its policies on land, infrastructure and finance in order to increase the access of low-income households to resources and enable them to improve their housing. In so doing, the government must become a partner instead of a patron. This is the framework within which the Federation deals with issues of housing for low-income groups in general, and housing finance in particular. The challenge is to establish a dialogue between the government and low-income groups and to do so in a way which works to the latter's advantage. The Federation's decision to initiate such a dialogue is based on both the need to inform the government of the situation of the low-income households and a recognition that the government is capable of delivering basic infrastructure and secure tenure to these groups. As if this is not enough, the Federation also aspires to factor-in gender because of an understanding that sustainable solutions are only possible with the involvement of women, who often take the lead in housing issues. The following section illustrates how the Federation between SPARC, Mahila Milan and the NSDF has attempted to meet these objectives.

THE FEDERATION: SPARC, MAHILA MILAN AND NSDF

A variety of grassroots organizations operate in low-income settlements in India, including NGOs and community-based organizations (CBOs) formed by low-income groups in order to fight against evictions or for service provision. Most NGOs, however, steer clear from the difficult issues of land security and basic amenities. Generally, NGOs 'adopt' specific slum settlements in order to provide primary health care, education and welfare and recreation services. Moreover, most NGOs and CBOs have worked in isolation from each other even though, as the case of SPARC, Mahila Milan and NSDF illustrates, they can come together to create a strategy which seeks to mediate between low-income households and government.

The alliance between SPARC, Mahila Milan and the NSDF

SPARC was set up in 1984 by professionals working on urban issues as a reaction to the perceived limitations of other NGOs who were focusing on service delivery without seeking to address the causal factors that impoverish low-income households (Mitlin 1997). The founders of SPARC sought to create partnerships with low-income communities in order to give them the confidence to undertake a dialogue with the government on issues of equity. SPARC began its work in a slum settlement in Mumbai called E Ward where large numbers of pavement-dwellers resided. It was argued that since pavement-dwellers were the poorest and most vulnerable group in the city, if SPARC could successfully work with them, then working with other better-off settlements would be easier.

It was while working in E Ward that SPARC came across a group called Mahila Milan ('Women Together') which had 600 women members. Mahila Milan's primary aim was, and is, to gain recognition for the vital roles women perform in low-income communities, to strengthen women's organizations and therefore give women a greater voice in both local and city forums. Through its work with pavement-dwellers and Mahila Milan, it became clear to SPARC that women were critical in the creation of housing for the family, the management of community services and the protection of homes from demolition. Ironically, however, when either NGOs or government officials visited low-income settlements, they sought out male leaders. Moreover, the socialization of women in low-income settlements was such that on the occasions of such visits, the communities 'presented' a male leader to speak on their behalf. SPARC believed that, since it was women who managed the housing process, it made sense to develop their skills and to encourage the formation of collectives that could participate in the upgrading processes.

SPARC's work in E Ward with the women pavement-dwellers subsequently attracted the attention of the National Slum Dwellers Federation (NSDF). Started in 1974, the NSDF consisted of a group of slum leaders from ten cities. Most of the leaders were men who had risen to such positions due to their active participation in organizing protests against slum evictions and demolition. By 1985-6 (when SPARC started working in E Ward), the NSDF wanted to transform its strategy and expand its operations away from leading protest marches and towards formulating options which would form the basis of a change in government policies towards slums. NSDF now wanted *direct* dialogue with the government, as they were dissatisfied with the way NGOs delivered services to their settlements. The next move was for SPARC and the NSDF to explore the possibilities of setting up a Federation in order to facilitate a dialogue with policy-makers and inform them about the wishes and aspirations of low-income households.

The Federation option

Starting in Mumbai, this Federation now works in twenty-one cities throughout India and has been instrumental in creating and sustaining communication links within and between cities, as well as internationally. It has cadres at community, city and national levels. Local federations in each city represent diverse groups ranging from pavement-dwellers to squatters on railway and airport land. Each local federation is divided according to location. Thus, the Federation has been instrumental in strengthening the institutions of the poor and providing them with a base from which they can initiate a dialogue with local and national government officials.

The Federation between SPARC, Mahila Milan and NSDF suited all three parties. For SPARC, the pool of leaders from NSDF and their outreach in the slums offered a channel to expand activities without necessarily having to expand its own operations. Prior to the creation of the Federation, SPARC, like other Third World NGOs, had found the process of scaling up its operations problematic. There was an insufficient pool of young and dedicated professionals who had the capacity to work as partners with low-income people. Moreover, the training costs of such an expansion were unsustainable. For the NSDF, access to the professional and managerial skills which SPARC had to offer was a necessary complement to their raw power. Both SPARC and the NSDF felt that the expansion of Mahila Milan from a women's collective in E Ward into a network working in partnership with the NSDF would be ideal. The leadership of the NSDF acknowledged that the slum-dwellers' movement had long used the anger of women to bring larger numbers of protesters onto the streets in rallies but that they had rarely sought to develop leadership skills among women or include women in their own leadership. The NSDF also acknowledged that women managed community resources and sustained community processes better than men and that with a clear collaborative strategy, men and women in communities could work together instead of competing with each other. Gradually, the realization has dawned that giving women access to positions of power and acknowledging the important role that women play in accessing housing is not an admission of defeat but rather the creation of new relationships between men and women. This, in turn, can result in increased access to resources for low-income communities.

The Federation has had an important impact on gender roles and gender relations within communities. With the support of NSDF, Mahila Milan has formed women's collectives to manage community resources including finance. By participating in the collectives, women have been emancipated and empowered. Interestingly, however, women in low-income settlements have often resisted emancipation when it has not included their families and communities. As a result, strategies that have sought to develop skills among women for their emancipation alone have rarely succeeded. Rather, strategies that begin by training women to access resources for their families and use increased self-esteem and social acceptance to improve family and community status have been more successful.

As the Federation has grown, so local organizations that were initially very dependent on a few key individuals are now successfully supported by a strong local leadership. In many cases, community members who have benefited from working with the Federation give something back by becoming trainers and participating in community exchanges. In this way, other groups in Mumbai and elsewhere in India are able to examine the solutions developed and adapt them to their own situation. Thus, the alliance between SPARC, NSDF and Mahila Milan has enabled the creation of a formidable mass of experience which the government and housing institutions are unable to ignore. This is illustrated by a recent agreement between slum-dwellers living on railway land and the railway administration which granted secure tenure to dwellers in return for them moving back from the railway line, thereby allowing trains to go faster and thus save money. Security of tenure has allowed the dwellers to upgrade their houses as well as gain access to basic infra-structure which they had lacked for the previous twenty-five years. Such a compromise would not have been possible if the communities had sought individual agreements with the powerful railway administration. Instead, the Federation created the conditions through negotiations with the government, who arbitrated between the railway and slum-dwellers until an agreement was reached. Moreover, this case has provided a precedent which the Federation has used to resolve similar problems in slums in other cities which are located on railway land.

FINANCE FOR HOUSING

Although the finance needs of households for income-generation have come a long way in the last two decades, the same cannot be said of housing finance. In the context of income-generation finance, global experiments have resulted in a change in the perceptions of communities, local and national governments and development practitioners to micro-finance (Hulme and Mosley 1996). In India, an increasing number of income-generating finance schemes have been initiated by central and state governments. These schemes are beginning to make micro-finance available to low-income men and women in both rural and urban areas. Of course, this is not to say that the problem of accessing this finance has been resolved. Rather, NGOs and low-income communities who want finance for income-generation purposes now know where to go, what bureaucratic rituals to expect and how to reproduce the organizational systems through which access to finance is made possible. There is now documentation available for new players to learn from the experiences of other finance managers and enough resources available to innovate, experiment and learn (Patel and Burra 1994).

However, the same progress has not been made in the field of housing finance either in developing countries in general or in India specifically. Finance for housing seems to be in an almost neo-natal phase of growth especially in the case of low-income groups (National Institute of Urban Affairs 1992, Mitlin 1996). A report by the Planning Commission of India (1992) found that although national housing finance requirements were close to US$ 1,400 million, the budgetary provision was a mere US$ 500 million. Only one-third of finance is provided by formal finance institutions and this falls to 24 per cent in the case of low-income households. Consequently, a high proportion of low-income men and women are excluded from formal finance and have to rely on informal mechanisms including savings, family assets and informal sector moneylenders (Patel and Burra 1994, Baken and Smets this volume).

The development of housing finance in India has been delayed for a variety of reasons. First, the government of India began to consider issues of housing finance rather late as compared to other developing countries and only initiated policies in its post-1987 housing reforms (Patel and Burra 1994). Second, housing has not been a priority area of NGO intervention: it is estimated that only four of the eighteen NGOs in Mumbai which are involved in both house upgrading and community finance provide housing finance (Mitlin 1997, Mehta 1994). This is partly attributable to the continued problem of illegal land tenure in many cities and the traditional role of NGOs as service delivery providers. As such, they are largely unable to grapple with the political implications of negotiating on the issue of land tenure. To complicate the situation further, there is a labyrinth of legislation which protects certain slums from being demolished but at the same time forbids residents from constructing permanent dwellings. Third, urban NGOs and community organizations working in low-income communities have found that household incomes fluctuate throughout the year, making scheme design difficult. Finally, housing finance for all social groups in India remains highly undeveloped. A study in 1995 revealed that 92 per cent of home-owners across all income groups did not use mortgages to construct/obtain their houses. An albeit flawed logic exists that if housing finance for the rich remains under-developed, one should not expect finance to be available for low-income households.

The Federation and housing finance for low-income households

It has been recognized that formal finance institutions cannot meet the needs of the urban low-income communities given the limited amount of capital that can be raised specifically for housing. This means that the actions of low-income households and community organizations are critical for accessing housing finance. To date, however, the importance of such community organizations has been largely unrecognized among government agencies (Patel and Burra 1994).

The Federation operates three kinds of savings schemes: crisis savings and credit schemes; income-generation schemes; and housing saving schemes. All three are interlinked in that most communities start with crisis credit schemes (which encourage members to begin saving) and as their confidence grows initiate income-generation and housing finance schemes. In addition, it should be noted that the availability of crisis and income-generation finance is important for housing construction as they encourage households to avoid using money put aside for housing to cover immediate crisis needs.

Crisis credit schemes are encouraged to include as many households as possible in low-income settlements. If all households in a large settlement wish to be involved, it is proposed that various small groups of 10–50 members are formed. Men and women who are interested in participating are trained and shown how the crisis finance funds work in other communities. Groups are encouraged to develop their own rules concerning the operation of the fund but the majority end up with broadly similar operating procedures. Within a period of three months, most communities are able to understand, agree and manage their own crisis fund. Within each crisis credit group, one woman is identified as a treasurer and savings groups are then linked together through the treasurers who are all members of Mahila Milan. Since most women are illiterate, many rely initially on their memory to keep track of loans and repayments but over time they keep written records of savings and loans by working upon their numeracy skills with school children. Most federations have a central meeting-place in which most financial transactions take place with savings being deposited and loans distributed.

Crisis credit schemes are often set up by the poorest women in a community who save very small amounts of money – sometimes as little as US\$ 0.03 per day. Even when the overall fund is as low as US\$ 5–6, women can begin to borrow small amounts for the purchase of medicine, bus tickets to find work or school books for their children. Although Mahila Milan and the NSDF develop the general parameters which bind the member co-operatives together, the decision of who gets the loans is left to individual collectives. Despite the fact that the amounts involved are very small, they fulfil crucial immediate needs while affording women members 'community acknowledgement' for having created these resources. Experience shows that the sums paid out as crisis credit are repaid very quickly and at least 95 per cent of loans are repaid in full (Patel and Burra 1994). Again, it is left to each collective to decide how to deal with defaulters. The Federation only intervenes if asked to do so.[2] Initially, groups varied on the amount of interest that was charged but now it is either 1 per cent or a flat service levy.

Savings are collected daily for two reasons. First, daily collections foster the habit of putting regular sums of money aside and, second, they create communication networks which allow leaders to develop a deeper insight into what is happening in each household. Often, these insights have revealed an impending crisis which can lead to group support for the household through a difficult period. This qualitative involvement is crucial to the Federation not only in terms of money transactions, but also in its organizational and mobilization processes.

Once crisis and credit groups have been set up, income-generating finance schemes can be organized. By 1996–7, over 50,000 households in Mumbai and a further 50,000 in four other cities were part of the income-generating savings groups. In total, these households represent 40 per cent of membership. Using the experience gained from managing crisis credit funds, the women in low-income settlements have begun to manage both internal and external finance lines. Capital for these loans is derived either from the saving pool of the Federation or modest revolving funds provided by grant institutions. The Rashtriya Mahila Kosh (a fund created by the Government of India to provide cheap finance to low-income women) is a vital external source that has extended an annual finance line of US$ 140,000 to Mahila Milan through SPARC. The fund offers finance for income-generating activities such as vegetable vending, carpet making, the recycling of garbage and the establishment of small provision stores. It is available at an interest rate of 8 per cent which the collectives then on-lend at a 'charge' of 12 per cent. Part of this charge is used to pay off interest with the rest constituting a compulsory savings component that creates a fund for short-term needs. The existence of this secondary fund means that members do not have to lie about what they want finance for as consumption finance is always available and can be kept separate from income-generation.

Housing finance is the third type of finance scheme offered through the Federation. The Federation's experience with housing finance is quite recent and to date only five co-operatives have extended housing finance to 700 households. With virtually no assured land tenure and no institutionalized practice of constructing housing for low-income groups, lending money to most households for housing remains problematic. However, the Federation is working to secure land tenure, standardize house construction and design costs and negotiate with housing finance institutions to convince them of the need to lend directly to low-income households. SPARC's specific role is to mediate between the decentralized system developed by the Federation and the vertical and central systems endemic to formal finance institutions. The main housing finance institutions with which SPARC has worked are the Housing Development Finance Corporation (HDFC, a private sector company) and the Housing and Urban Development Corporation (HUDCO, a public sector finance company). These companies receive special funds through bilateral development assistance and are experimenting on ways to reach low-income communities. Both have tried to involve NGOs in the delivery of funds but the small number of NGOs working in this sector and the procedures which both HDFC and HUDCO employ make collaboration difficult.[3]

In the meantime, the Federation is assisting communities to begin saving specifically for housing so that they can demonstrate their commitment to jittery bankers by the time they are granted secure land tenure. A savings record also serves to reduce the amount of money households have to borrow. Two types of housing loans are extended: the first is for incremental improvements to housing and the second is to co-operatives which have acquired land on which to construct housing. For incremental improvements, loans of up to US$ 200 are extended to individuals who belong to collectives which operate along the same lines as income-generating groups (Mitlin 1997).[4] Initially, these loans were part of the revolving savings pool but now that HUDCO focuses specifically on housing, a special fund (which operates along similar lines to the Rashtriya Mahila Kosh) has been set up.

The second type of housing loan is available to co-operatives which have acquired land but which are often unprepared to begin construction. Here, an interesting exercise is undertaken whereby the Federation advocates the pooling of savings which are then guaranteed by the Federation. This serves to create an external creditworthiness through the good offices of the Federation. One such co-operative is the Markandeya Housing Co-operative in Mumbai, which was formed in 1989 and consists of ninety-five households. Markandeya illustrates the problems that many housing co-operatives have faced. The co-operative was able to gain access to allotted land in Dharvi on which it planned to build walk-up tenements at a total estimated cost of US$ 100,000. It was agreed that HUDCO would provide 65 per cent of the finance, 20 per cent would be made available through a government grant and the remainder would be obtained from the co-operative members. However, it took three years to obtain the lease for the land from Mumbai Municipal Corporation, during which time no finance was available. In order to obtain finance, SPARC had to provide a bank guarantee, which it was finally able to do through an international NGO, the People's Low Cost Housing Program in Asia (SELAVIP). Further and more significant delays occurred due to obstacles in releasing finance which led to a significant increase in costs following rapid inflation in the construction industry and the 'price' of securing numerous building consents. The result of these delays has been that as many as 20 per cent of the original ninety-five households can no longer afford the housing, the building of which began only in 1997. These households have been given the option of selling their units back to the co-operative at the market value and then joining other co-operatives who are just starting the process (Mitlin 1997).

The Federation's experience with housing finance remains limited and there are many issues that still need to be addressed. The most essential factor is the vulnerability of low-income groups who with no insurance or support can default on repayments due to illness or unemployment. Consequently, flexibility in setting up repayment schedules is critical, as SPARC's own experience has shown; when it first insisted on regular repayments on its own loans, communities organized themselves so that households swapped repayments as resources allowed. Now, the Federation encourages communities that are saving for housing to swap finance between branches of the Federation if the trust between the two parties is sufficient.

CONCLUSION

Time is a critical factor when reviewing progress in the field of housing. This has presented challenges for community organizations and activists to sustain the interest and involvement of large numbers of households while the housing process moves at a snail's pace. It is crucial that this problem is understood in an environment of 'instant karma' and quick-fix solutions which seem to infect most development interventions.

This chapter has argued that the creation of strong community organizations capable of negotiating with the government or finance institutions is crucial to gaining more resources for low-income households. The Federation option provides the potential for low-income communities to work with city authorities to make a difference in housing and poverty reduction. This work is made both more powerful and complex due to the Federation's inherent preoccupation with the roles that women perform in these activities. It pursues educational and organizational strategies which balance issues of gender within this context. Hence, we constantly reiterate that women are central to the processes which involve poor communities. They hold in their hands the ingredients which are essential for sustainable solutions to the problems of the poor, and although they cannot provide the resources to solve the problems, their capacity to design these solutions, manage the strategy and judge whether the outcome is viable makes their participation crucial.

The finance schemes described above create the conditions to ensure that women are central to and remain in charge of the finance process by maintaining the collective strength of the Federation network. Having started at a small scale, women get used to working with larger amounts of money. Additionally, as more and more communities participate and refine this decentralized, accountable and transparent process, they become attractive organizations to lend to as they absorb much of the administration costs which financial institutions would otherwise have to sustain themselves. In return, communities can (hopefully) negotiate preferential interest rates.

However, the model that the Federation has developed is vulnerable at four distinct points. First, the entire process depends on trust. It works because Federation members trust each other to act in the interest of the collective good. Second, to date, the process has been very decentralized and this has been critical to innovation and adaptation. However, it is debatable whether such decentralized activities can be maintained as the process grows. Third, a delicate balance has to be maintained in relations with the government, which would like to absorb SPARC by focusing all its activities on government initiatives. Few institutions can maintain a co-operative relationship with the government without either becoming sub-contractors or entering into conflict with it. In order to maintain a healthy distance from the government, SPARC and the Federation are continuing to provide new alternatives for housing and finance. Fourth, is the question of the sustainability of the Federation model. SPARC has never attempted to build a development model that is financially sustainable without the continuation of external support. In fact, until now the more important question for SPARC has been whether the process is growing and therefore attracting resources. Despite this, costs are kept at a minimum and the Federation is built upon the principal of voluntarism with only a small core of 35–40 full-time staff.

To SPARC the issue is really who is willing to experiment, innovate and create systems and institutional arrangements that suit low-income groups? It seems clear so far that neither the government nor financial institutions are willing to take on this role. Moreover, although national, bilateral and multi-lateral development organizations have appropriated the language of 'participation' in policy papers, their increasingly centralized and vertical structures impede rather than facilitate the delivery of resources. SPARC's experience with HUDCO and the HDFC demonstrates the limitations of working with formal finance institutions which want to integrate low-income groups into inflexible financial systems with loan structures similar to those used by higher-income households. Formal finance institutions often push NGOs to take responsibility for financial accountability and monitoring even when it is clear that many do not have the professional resources or in some cases the capability of carrying out these tasks. As a result, the onus is left on low-income communities and their allies to use small amounts of subsidized finance to experiment, make mistakes, pick up the pieces and forge ahead to explore new possibilities.

NOTES

1 India has a federal political structure where the central government provides directives while state governments develop policies and finance them.
2 The unique feature of this process is that while decision-making about loans, savings and transactions is decentralized, it is possible to use collective strength when attempting to obtain finance and weighing risks.
3 Perhaps the best example of this is HUDCO's rules that did not allow co-operatives to access housing finance by themselves. However, in collaboration with SPARC, this rule was subsequently amended.
4 The Federation encourages those individuals who want to upgrade their dwellings to form co-operatives on the grounds that this enables them to access better building materials as well as giving them the chance to build collectively.

11

GENDER AND MICRO-FINANCE IN SRI LANKA

The experience of the Women's Credit Union[1]

Alana Albee and Nandasiri Gamage
Freelance consultant, UK and community activist, Sri Lanka

INTRODUCTION

According to official statistics, Sri Lanka has had an impressive record of development with only 4 per cent of children dying before their fifth birthday, a high average life expectancy of seventy years and 85 per cent of women deemed to be literate. However, during the 1980s and 1990s, conditions have improved more slowly or not at all: unemployment exceeds 20 per cent of the labour force; prices have multiplied five-fold between 1977 and 1990 and government expenditure on social welfare has declined to pre-1960s levels. These conditions have had a devastating impact on low-income households, especially women, as increased poverty has forced them into the workforce as either co-, primary or sole income earners.

The extent of urban poverty is exemplified by the case of Colombo where there are approximately 850 low-income settlements with a total of more than 350,000 residents. The endurance of poverty here has led many people to question the whole concept of development. As argued by one of the authors, Nandasiri Gamage:

> Sri Lanka is classified as a 'developing' country. My father read about this as a young man, now I read about it in my middle-age and my son reads about it in school. Yet, it is unclear to us what we, as Sri Lankans, want to achieve and how we want to define 'development'. Are the huge buildings and roads and the tourists on the seaside beaches indicators of 'development'? Whose needs will be met by these?
>
> (Albee and Gamage 1996: 39)

Concerns about the failure of macro-development efforts to alleviate poverty have resulted in greater attention being paid to alternative approaches which enable low-income men and women to become active agents of their own development.

This chapter argues that this can be done through supporting low-income groups to gather information about their circumstances and resources, to analyse their situation, prioritize the actions they wish to pursue and work out the means of implementing these actions. Savings and credit groups are an important part of such a micro-development approach focused on the needs of low-income groups. Within urban centres, although inadequate incomes are a major cause of poverty, it is the lack of assets which underlies the vulnerability of low-income men and women to economic shocks (Chambers 1995, Pryer 1993). Without community-based insurance and informal finance schemes, these households are often forced into debt. Savings and credit groups build upon the capacity of low-income people to save and offer them a viable alternative to own and control resources.[2] Access to informal finance is especially important to low-income women, who may face particular difficulties in gaining access to income, resources and services. Moreover, they can be a particularly valuable and empowering tool for women as not only do they offer a reprieve from the constant worries that accompany subsistence, but they also serve to fund the productive activities sustaining the family.

This chapter concentrates on the experiences of the Colombo Women's Credit Union (WCU) which was formed in 1989 with government, United Nations' Centre for Human Settlements (UNCHS) and Children's Fund (UNICEF) assistance. The project, initially operating from the government's National Housing Development Authority, offered its members a mechanism for saving and making loans.[3] In 1992, the WCU became an independent organization managed by low-income women and therefore illustrated that women could act as agents of development. The WCU has a very clear and strong approach to gender and challenges the thinking that access to finance alone leads to empowerment. It recognizes the importance of low-income women's control and management of financial systems not only at the household level but also in the community. Consequently, there has been a concerted effort to make low-income women responsible for decisions on who obtains loans as well as who controls these funds. In so doing, the WCU goes beyond a 'banking *for* the poor' approach to a 'banking *by* the poor' approach.

Loans obtained from the WCU are used for a variety of purposes including housing construction and improvement. According to Sirivardana (1994), the government of Sri Lanka has attempted to ensure that improved housing is integral to its vision of poverty alleviation. This commitment is borne out by the initiation of the Hundred Thousand House Programme and, later, the Million Houses Programme (Sirivardana 1986). It has implemented a strategy of supporting an incremental development process in low-income settlements and of working with local residents through a joint planning process. For people living in low-income urban areas, gaining access to land, common amenities and housing is often a main priority. Indeed, mobilization for housing has proven to be a successful foundation from which community members gain experience at articulating their needs, identifying resources and managing the development process.

This chapter will begin by outlining the structure and organization of the Women's Credit Union before moving on to examine its savings and lending programmes. It will then examine the provision and use of housing loans. Using examples from the field, it will illustrate that in spite of best intentions, there are points where the WCU has been vulnerable. The chapter will conclude by looking at the future challenges to the WCU such as the issue of scaling up its activities. Throughout, the chapter illustrates the importance of low-income women in creating, managing and accessing the credit union's resources.

THE WOMEN'S CREDIT UNION (WCU)[4]

The WCU (known formally as the Women's Thrift and Credit Co-operative Society) has evolved over several years. While it adopted some of the features of other organizations such as the Grameen Bank and Thrift and Credit Co-operative Societies, the WCU purposely avoided transferring a specific 'model' of finance to Sri Lanka. Instead, low-income women were encouraged to build an organization based on their own needs and constraints. In so doing, the twin objectives of developing a mechanism for savings and making loans and expanding women's capacity-building were achieved.

The formation of the WCU is based on certain key principles. These include, first, a decentralized approach to credit which contrasts with the approach of formal finance institutions where decision-making and management are hierarchical. The approach taken by the WCU guarantees that low-income men and women manage and do not simply 'receive' resources. Initially, the WCU supported the formation of small pre-co-operative groups, but by 1990, there was an appreciation that these small credit groups were unlikely to be sustainable in the long term, a fact borne out by the collapse of two of the original groups. A strategy of 'federating' was, therefore, seen as a means through which groups could gain strength, share their experiences and solve problems. Moreover, this strategy would enable members to discuss the nature of the lending system and to develop ways of increasing its effectiveness.

Second, the habit of saving is promoted *before* any loans are issued; which enables the development of a self-reliant base of lending capital. Third, the common bond between members (who usually live or work in the same location) is emphasized and shares are sold to members to promote the concept of common ownership.[5] Finally, equity is assured by basing voting rights on

WCU membership as opposed to share ownership. There are now eight branches of the WCU in the low-income areas of Colombo and a further twelve in rural areas which collectively reach 3,600 women.

At present, the WCU is supported by two organizations, the Praja Sahayaka Sewaya (PSS) and the Kantha Sahayaka Sewaya (KSS). The PSS was set up in 1990, to serve as an umbrella organization under which various activities, from housing co-operatives to the production of a local newspaper, could be undertaken. The KSS works under the PSS as a support organization to the WCU. Its members work part-time as social mobilizers by meeting potential members and facilitating the rapid expansion of the credit union. As all KSS staff are from low-income areas, they are able to explain the process of credit union formation and federation in terms which can be easily understood by other women.

Although wishing to avoid operating through a rigid hierarchy, the WCU has found it necessary to decentralize its support activities and, from 1994, to form a zone level to enable learning and sharing of experiences to take place. It supports the formation of groups which can subsequently join to form primary branches. Groups are formed in an *ad hoc* manner as information spreads by word of mouth (as women talk to their neighbours, in the market and at almost any meeting-point including health clinics, water points, on the bus and as they travel on the train). Many women subsequently contact the KSS support workers in search of more detailed information and for advice on the formation of groups.

Initial contacts with local communities pass on practical information as well as discussing wider issues such as the purpose of the WCU, the function of groups, the procedure of meetings, the way in which groups should choose their leaders and discussions of social development activities. This is the beginning of a process of broadening low-income women's understanding of societal issues and the power gained through unity. It is also a dynamic time of building women's self-confidence, inner strength and dignity, as well as encouraging others to join their own development process. Experience has shown that women frequently require a minimum period of six to eight weeks to decide whether or not to join a group. The pressures of funding and reporting deadlines which so often force the pace of government development projects, however, are not factors which determine the growth of the WCU. On the contrary, it is recognized that it is important to give women adequate time to form groups so that there is trust among members.

Once formed, most groups consist of between five and fifteen members, usually women.[6] Members of a group need not produce or sell the same type of product, but must live in close proximity to one another and have a common trust. Basic group rules are that those attending group meetings should only be low-income women,[7] only one member from each family should be in any group and no moneylenders or loan-sharks should join. Additionally, the leaders of any group should not be actively involved in politics, be office-bearers in community development or other associations, nor be involved in other development support teams (such as the PSS). Each individual must contribute US$ 2.50 to join the WCU. This is composed of US$ 2 for the allocation of one share in the Union, US$ 0.20 as a membership fee, a US$ 0.20 monthly contribution to the welfare fund and US$ 0.10 as a monthly savings contribution.

Primary branches are the structural representation of group federation and provide services and monitor the work of groups. Each branch is made up of approximately eight groups and, by 1994, a total of eighteen primary branches had been established. The branches operate as mini-banks on a profit-loss basis and aim to be financially viable by carefully pooling and circulating the resources of the member groups. As accumulated savings equal, or indeed exceed, accumulated borrowing, the branches are usually solvent from the initial stages. Consequently, the groups which make up the primary branches risk losing their own savings if they default on loan repayments.

All funds, except compulsory and voluntary savings (see below), are managed at the primary branch level. Groups apply to their primary branch for lending capital seven days in advance of requirements. In addition to providing capital for the groups, primary branches balance the cash requirements of their borrowing from the KSS/PSS and serve as a repository for savings funds, including the compulsory individual saving of US$ 0.10 per month as well as a number of other savings mechanisms devised by the WCU (common and group welfare funds, voluntary savings, fixed-term deposits, children's savings deposits and non-member savings). The branches are managed by the treasurers from each group who must have practical experience in handling funds at the group level (for a minimum period of five months) before they manage the larger and more complex systems of the primary branches.

Having detailed the organization of the WCU, the following section examines the savings programmes of the credit union.

THE PROMOTION OF SAVINGS

Savings by group members form the core capital of the WCU and the obligatory nature of savings is one of the key factors which ensures that the WCU is owned and controlled by low-income groups. Urban women members prefer the discipline and routine of weekly deposits of savings which they find easier than trying to save by themselves at home. The habit of saving is fostered through the use of promotional pamphlets devised by the KSS and PSS to stimulate group discussion. These pamphlets are a practical tool to help women to understand how savings can accumulate substantially over time even when only a small amount can be regularly deposited. For example, one pamphlet details the amount of money that could be saved if one member of that household stopped smoking.

Various types of savings have been developed by the WCU over a four-year period and are introduced gradually to each group. Some programmes are compulsory in nature while others are voluntary. Compulsory savings form the capital base from which loans are drawn for allocation to individual members and groups. An annual interest of 15 per cent is paid on the deposit of compulsory savings. In the case of individuals, monthly savings of US$ 0.40 are managed by the group as an emergency fund which can make loans for the purchase of medicine and to cover unexpected household expenses. Group compulsory savings amount to US$ 0.10 per month per member, and are managed by the primary branches. In addition, a common welfare fund which is supported by a monthly contribution of US$ 0.20 per month per member, issues loans to cover funeral expenses.

A second type of savings mechanism is voluntary savings and fixed-term deposits. In the case of the former, any member may save additional money at the group level for her children or herself but receive no interest for doing so. Furthermore, an individual may make fixed-term deposits, whether or not they are a member of the WCU and receive interest on savings (15.5 per cent) of six months and one year. Every subsequent six-month period earns interest of an additional 0.5 per cent up to a maximum of 18 per cent.

The groups are organized in such a way that members who save together go on to borrow together. This reduces the likelihood of the marginalization of an individual member, as we will see below.

ISSUING LOANS

Before lending money, primary branches have to ensure that there is an adequate level of share capital within a group (a 1:10 share/capital ratio must be maintained at all times), that welfare funds and compulsory savings contributions have been made on time and to the correct amount, that there has been full attendance of group and branch level meetings, and that past loans have been utilized for the agreed purpose. The primary branch subsequently issues loan capital to each group which in turn extends loans to their individual members. A group is never issued overlapping or simultaneous loans from the primary branch, just as no individual borrower can receive two loans at the same time.

The provision of loans is divided into four stages. Groups are eligible for stage one loans to cover emergency expenditure after about five months of compulsory saving. These loans are typically small (maximum US$ 2); they incur a monthly 'service charge' of 4 per cent to be repaid by a date decided upon by the group.[8] Loans at the second stage range from US$ 5 to US$ 10 and are used to fund both small-scale productive activities and consumption needs. They are usually repaid within a month. After saving for eighteen months, groups can move on to the third stage when they can apply for loans of US$ 20–40. These loans are paid back within two to four months and groups incur a 4 per cent service charge per month. In the last stage, groups can borrow up to a maximum of US$ 300 per member: loans of US$ 200 are commonly extended for income-generating purposes and loans of US$ 300 are available for housing. The smaller loans incur a monthly service charge of 2.5 per cent, while the larger loans incur a 2 per cent service charge per month with the condition that a minimum of US$ 5 has to be repaid every month.

This approach of graduated loan sizes has two benefits. First, it enables women to build their confidence through the experience of managing small loans before moving on to larger commitments. Second, it provides a built-in incentive of offering increased loans while ensuring repayment through the inter-guarantee system between borrowers: if any group fails to repay the primary branch at any stage all members become ineligible for larger loans. An important feature of this process of graduated loans is that each woman progresses through the various stages with her group. Thus, at any given time all group members have the same loan size and payments due. Only at stage four does this pattern alter when some women within a group can access larger housing, income-generating or ceremonial loans while the others wait their turn (commonly determined on the basis of urgency of need).

The concern of individuals to retain their group's credibility and internal trust, the loss of accumulated and children's savings, and access to death benefits if loans from the primary branch are not repaid on time ensures that default at both the group and primary branch levels is virtually non-existent. In the rare cases where loans are not repaid on time, the branch is compelled to fine group members. Delays of one day result in an additional charge of 1 per cent, a delay of not more than fourteen days leads to an additional two weeks' charge on the outstanding balance and a delay of more than fourteen days results in an additional one month's charge. If a group is fined twice during any stage of borrowing, it is downgraded to the previous loan size. Here, the idea is that two delays are an indication that borrowers are unable to bear the burden of repaying such large loans. When instalments from individual members to the group are late, the individual is fined. These fines are collected by the group and distributed among the other members at the end of the year.

The importance of these loans for housing improvement and construction is considerable as housing is a main priority for WCU members. It is to this topic that we will now turn.

HOUSING LOANS

In areas where land rights have been acquired and service upgrading has taken place, there is a strong motivation and interest among women to form savings and credit groups in order to obtain housing finance. Housing loans were first extended by the WCU in 1993 following floods in Colombo which displaced significant numbers of low-income households. Initially ten loans were given to women flood 'refugees' who had temporarily settled with their families in the offices of the PSS and KSS in Seevalipura. These women were unable to access government housing loans because of guarantor requirements and the absence of earnings certificates. With capital made available from the UNCHS through the PSS, the KSS supported the formation of two 'special' groups for housing loans and stressed that future access by other low-income people to housing loans would depend upon their repayment performance.

Loans of US\$ 200 were issued first and then, depending on the progress of individual house construction, further loans were issued until the house was complete. The key rules which guided lending were that any woman's outstanding loan balance would never exceed US\$ 300 and that only those women living in areas where they had obtained land rights (security of tenure) were eligible for housing loans; squatters were not eligible.

Women give several reasons for their interest in housing loans including the link between an upgraded, improved house and a successful business. The desire to improve the design of housing is often linked with the need to increase the production space needed for home-based businesses. For example, those women involved in food production are concerned about the design of their kitchen and express a need for greater floor space, the provision of ventilation and enlarged counters. Such improvements in house design do have a beneficial impact upon business performance as seen by the growth in small shops. Less visible home-based producers, most of whom are women making food and consumables, also increase. Once their homes have been upgraded, women and their families devote an increased amount of time to their enterprise. As one woman leader explained: 'Amongst the poor there are standards. People with upgraded and clean houses have more customers and better businesses!'

Following the successful repayment of housing loans in Seevalipura, the system was extended to other members of the WCU. Selection has been carried out on the basis of being a member of a long-standing group which has managed US\$ 40 productive loans for an extended period of time. As of late 1994, five out of the eighteen primary branches had issued housing loans to approximately fifty women.

WHEN THINGS GO WRONG

No credit and savings operation is trouble-free and there have been cases when money has gone missing, not been repaid or repayment has had to be rescheduled. Although these are infrequent occurrences, the following cases illustrate the problems which have arisen and how they have been resolved by the WCU with assistance from the KSS and PSS.

Case 1: Personal emergencies

The first case of money being stolen occurred in 1990. A member from a group that had been issued US$ 70 by the primary branch and had savings in their cash box visited the PSS office and said 'we don't have money in our box for emergency loans. The group leader told me to come and tell you.' The same week the group met with the PSS, and checked the records and the cash box. There was no money in the box, although the written records showed there should have been. The leader admitted taking the money for her ill child and her husband's medical expenses. The group wanted to continue the lending system but they had to revise their leadership. The old leader was allowed to remain in the group as a member but was stripped of her office-bearing position. The group began borrowing again but the former leader's situation meant she was not a reliable borrower and it became obvious that she was holding back her group's progress. Eventually she was asked to leave the first group and join a new group which was at the first loan stage. The new group is fully aware of her previous difficulties and sympathize with the difficult circumstances.

Case 2: The stolen box

In a very poor area of Colombo near the Kelani River where people live in homes made of straw, a group leader left her home to take her son to school. When she returned she discovered that the cash box containing the accumulated savings of the group was missing. The leader went immediately to the KSS office to explain as well as to the police station. She suspected a particular man whom the police arrested but subsequently released when the suspect's wife came to the house of the group leader and forced her to withdraw her suspicions against him. Eventually, the leader went to the KSS office where a new box was issued with a US$ 20 loan. The money was given to cover the costs which she had to bear in getting to the KSS and police to report the incident. None the less, the group leader had to repay the stolen money monthly at 1 per cent interest, which took her ten months.

Case 3: The key holder

In the 'Kumari' credit group, the cash box was kept by the treasurer whose husband was a heroin addict. From time to time, the treasurer would request the group member who kept the key to give it to her. The keyholder was not an active woman and was known for frequently giving the key to other group members too so that they could withdraw money themselves. The general rule for all groups is that every week at the group meeting the cash box has to be opened and the money counted. But this group didn't do this regularly and some of the money (about US$ 80) disappeared in late 1993. The treasurer had taken the money, but it was nearly six months before the other group members realized the problem when she admitted taking the money. The group members met with the KSS and discussed the problem. The reliable members formed a new group which excluded the treasurer and took the responsibility for re-collecting the outstanding money from the treasurer's father. By October 1994, US$ 60 had been repaid to the newly formed group.

Case 4: A group collapse
One well-established group collapsed at the final loan stage in which each member could borrow US$ 40. Two members of this group of eleven did not repay any of their loan because of a disagreement over leadership: some group members felt the leader was unable to keep the accounts and meeting minutes adequately. One member who was especially cunning wanted the leadership and frequently went to the KSS office to complain about the existing leader. The KSS investigated and found that the leader had power and influence over others in the community. When the group eventually collapsed, its accumulated savings and shares exceeded the balance of the outstanding loan instalments to be paid.

These 'failures' have resulted in important changes which make future savings and lending programmes undertaken by the groups less vulnerable. For example, as a direct result of the first case, the overall system for the use of the cash box at the group level was revised. Now no single person is allowed to keep both the money box and the key. Therefore, to remove any money without the group's approval requires a conspiracy of more than one member.

TOWARDS SUSTAINABILITY AND SCALING UP

One of the significant challenges facing the WCU at present is the issue of scaling up its activities. Although the WCU believes that maintaining the levels of participation, control and management by low-income groups is vital to its successful operation, it is aware that as of 1994, it's 3,500 members represented less than 2 per cent of Colombo's low-income women. With a view to expanding the membership, the union has begun to consolidate and look to scaling up its operations. However, the issue of scaling up needs to be put in perspective: to those who have worked to bring the WCU into existence, having 3,500 members of whom more than 2,000 attend their annual general meeting is a remarkable accomplishment for a low-income people's initiative which is not led by a political party or an international NGO.

None the less, the KSS and PSS are beginning to challenge their own perception of scale. One critical issue which needs to be addressed in the process of scaling up is whether the union should maintain existing lending mechanisms or develop new ones. This involves either increasing the number of loan types and savings mechanisms for the benefit of established members or stabilizing mechanisms in ways which give future emphasis to encouraging new members. This issue is not easily resolved as the strongest pressure is exerted by women who are already members and would like the benefits of larger loans. In turn, this draws into question the commitment of existing members to focusing the credit union's energies on bringing in more members.

Although unresolved at the time of writing, it is likely that a compromise will be reached which provides some form of new loans to existing members as well as reaching new borrowers.

Although the union is working toward financial sustainability, there are key points when a primary branch requires additional capital. The main difficulty is whether a critical mass of borrowers can be reached to generate the required income to cover administrative costs as well as the expanded demand for capital when several groups reach the stage of being eligible for the large housing loans. Previously, capital shortages were tackled by either postponing a group's access to a primary branch loan for a short period while further repayments and savings were accumulated or by seeking additional capital through a loan to the branch from the KSS or PSS. As more groups join and 'graduate' to make demands upon the system, such an approach may be unsustainable.

CONCLUSION

The WCU in Sri Lanka reinforces some of the well-established principles which the global co-operative movement has advocated for over 150 years. However, it has also added a number of innovative features of its own such as the formation of pre-co-operative groups, graduated loan sizes and the organization of group solidarity. A fundamental aspect of this credit union's success has been its firm commitment to low-income people, especially women, taking control of resources. This commitment goes beyond the mechanistic notion of development as increasing physical assets and the flow of goods and services. People mobilizing, inquiring, deciding and taking action of their own is an end in itself and not just a means of development. If low-income people are moving in this way to assert themselves, who is to say they should be 'developing' differently? As we have stated elsewhere:

> The communities whose efforts are described here may be 'low-income' by material standards of the so-called 'rich', but are immensely rich themselves in culture and values. This is illustrated in the collective endeavour of the credit union. We are working towards better standards of living through credit and savings, but our development is also about our strength as a union and a voice for the poor.
> (Albee and Gamage 1996: 36)

The challenge for external organizations is to support the WCU and the agenda and priorities set by low-income people themselves. External organizations often utilize such initiatives to deliver and address their own priorities. This reverses the process of developing low-income people's self-determination but is often an inviting 'threat' to low-income men and women who are lured by resources and promises of partnerships which only prove to reinforce dependency.

An important message lies in the choice to use a credit union mechanism as a tool for development. However, hitherto, the wider issues of power and control have been inadequately considered. If 'self-determination' and the empowerment of low-income people are 'goals' then careful consideration must be given to the type of organizational structure through which credit and savings are promoted. It is essential to address the issue of who controls and holds power. Structures which hold the power for, and on behalf of, low-income men and women promote dependency rather than self-determination.

Credit unions, such as the WCU, which are democratically owned co-operative societies and are built on principles of membership and equity address the questions of power and control. Their foundation is made of small pre-co-operative groups that federate together to form the 'union' which is not controlled or owned by non-borrowers. This challenges the more conventional 'banking for the poor' approach in which institutions are built by professionals who live a life very different from the low-income borrowers whom they seek to serve. The current challenge is to move beyond the delivery of credit and savings to a situation in which low-income groups own the process as well as the mechanisms of development. This challenge is beyond both welfarism and the free market and requires the recognition of people's abilities to create their own organizations.

NOTES

1 This chapter is a substantially edited and revised version of an earlier publication entitled *Our Money Our Movement: Building a Poor People's Credit Union* (1996), Intermediate Technology Publications.

2 Although credit unions may charge a higher rate of interest than banks (for example in Sri Lanka, the Women's Credit Union charges 2 per cent per month as opposed to the 21 per cent per annum charged by banks), these unions still provide an essential service. With no recourse to banks and in the absence of credit unions, low-income people have had to rely on moneylenders who may charge interest rates of up to 240 per cent per annum.

3 The WCU concentrates almost entirely on women because it recognizes that income earned by women is more likely to benefit the family and children.

4 Figures have been converted to US dollars at the 1994 rate of US$ 1 = Rs. 50.

5 Shares are sold for US$ 2.00 each and it is mandatory that each member own one share for every US$ 20 borrowed. For example, each woman must have purchased two shares before she can borrow US$ 30.

6 Groups do sometimes contain men at the request of women members. However, in 1994, the groups in Colombo had no male members while some rural groups, such as the one in Kalutara, contained only a handful of men. Office-bearing posts are never held by men and in most cases, they participate 'quietly' but positively in the group.

7 Indeed, comparatively better-off women living in low-income settlements seldom join these groups because of the length of time they have to save before receiving a loan and the comparatively small size of initial loans. As explained by one KSS worker: 'The better-off are not patient enough to come together, save and eventually after a minimum of five months, receive a small loan.'

8 Interest paid on borrowed money is referred to as a 'service charge' as this is considered to be a simpler and clearer concept for borrowers.

12

INFORMAL FINANCE AND WOMEN'S SURVIVAL STRATEGIES

Case studies from Cameroon and Zambia

Ann Gordon
Natural Resources Institute, UK

INTRODUCTION

Throughout most of sub-Saharan Africa there is no indication of decline in the rate of urban growth and it is estimated that cities will grow, in the current decade, by 10.5 million people or 5.3 per cent per annum (UNDP 1992). In recent decades, this urban growth has consistently surpassed economic performance. Moreover, economic reforms have resulted in a significant reduction in formal sector employment and real wages, while cuts in public expenditure on education, health services and food subsidies have intensified the poverty and vulnerability of urban low-income households. In an attempt to maintain incomes, households have engaged in a variety of survival strategies including switching from wage to non-wage employment in the informal sector and the increased participation of women in the labour force (World Bank 1995). Household composition has also undergone change as extended households have become more common allowing members to pool income and resources in an attempt to minimize risks. The incidence of both *de facto* and *de jure* female-headed households has also increased.

These strategies have had a particular impact on gender roles and gender relations within households and in wider society. Women's increased participation in the workforce, and particularly in the informal sector, has been caused by declines in both the formal sector and their traditional marginalization by this sector. The latter is partly attributable to the fact that women generally have fewer skills and lower levels of education than their male counterparts. As a result of this discrimination in the formal sector, women may be obliged to move more rapidly into the informal sector for which surveys typically show higher rates of women's participation (Mills and Sahn 1996). The 1987 census in Cameroon, for example, reveals that 57 per cent of urban women work in the informal sector as opposed to 39 per cent of men.[1] Recent research in Guinea suggests that among retrenched public-sector employees, women are likely to face a longer duration of unemployment in the wage sector than men and a shorter duration in the non-wage sector (Mills and Sahn 1996). Moreover, the pressure on women to seek employment is also the result of declines in male employment so that women may find themselves more likely to have partners who contribute less (or nothing) to household income. This means that women have to assume primary responsibility for feeding the household – especially children.

This chapter draws upon research by the author in Cameroon and Zambia for the Natural Resources Institute (NRI).[2] This work sought to investigate the survival strategies undertaken by low-income women to ensure the survival of their households during periods of intense economic decline. It uses the concept of 'livelihoods systems' as a framework for looking at these survival strategies. As noted by Grown and Sebstad (1989: 941–2), a livelihoods system:

> refers to the mix of individual and household survival strategies, developed over a given period of time, that seeks to mobilize available resources and opportunities. Resources can be physical assets such as property, human assets such as time and skills, social assets, and collective assets like common property (forests) or public sector entitlements. Opportunities include kin and friendship networks, institutional mechanisms, organizational and group membership, and partnership relations. The mix of livelihood strategies thus includes labour market involvement; savings, accumulation and investment; borrowing; innovation and adaptation of different technologies for production; social networking; changes in consumption patterns; and income, labour and asset pooling.

In Cameroon and Zambia, urban women's livelihood systems concentrate upon small-scale income-generating activities, membership of informal finance groups and personal networks in both urban and rural areas. Informal finance groups are particularly important as they complement other strategies by permitting access to sizeable funds for investment, consumption and crisis needs.

Fieldwork was conducted in Cameroon during 1993 in Mabanda, an area of spontaneous settlement originally on the outskirts of Douala, but now engulfed by the expanding city. The fieldwork involved meetings with key local figures, 'transects' in the community including market visits and interviews with community members, group meetings with women to discuss their activities, concerns and constraints, and a wealth-ranking exercise. In addition, follow-up interviews were conducted with selected women for the primary purpose of understanding the survival strategies adopted by them.

Fieldwork in Zambia was carried out in 1994 in Lusaka and Livingston in collaboration with an NGO, CARE-Zambia, which was implementing the Project for Urban Self-Help (PUSH) in low-income compounds. This research was undertaken as the project moved from its initial phase of infrastructure improvements to an emphasis on participatory planning and implementation of project activities by residents themselves. The decision to work with CARE was based on a number of factors including CARE's existing contacts with the target group, its experience with the women and potential to follow up on the work. More than twenty group discussions were held in addition to fifty in-depth interviews with selected women and transects in local communities. During both field studies, the emphasis was on a qualitative understanding of women's survival strategies and their perceptions of constraints and opportunities. Most of the interviews were with women and men were rarely present.

THE SURVIVAL STRATEGIES OF LOW-INCOME WOMEN IN CAMEROON

Before identifying the survival strategies adopted by women in the Cameroon, it is necessary to outline the context within which these have developed. Economic growth accelerated in the 1970s following the discovery of oil to average 11.5 per cent per annum in 1976–81 and 5.9 per cent per annum in 1982–5 (Derrick 1992). Concurrently, the public sector doubled in size and employment opportunities in Douala and Yaounde expanded in both the commercial and public sectors. In 1995, 45 per cent of Cameroon's population was estimated to be urban (World Bank 1997).[3]

Even when the economy of the country started to decline in the late 1980s, its GNP per capita still stood at just under US$ 1,000, more than twice the sub-Saharan average. However, a combination of domestic and external factors including falling oil and agricultural export revenues, increased public spending, capital flight and doubling of the external debt have led to a budget and external payments crisis resulting in the adoption of economic reform measures. One result of these measures has been a marked decline in formal sector employment in both the public and private sectors. Although the direct impact of structural adjustment on poverty in Cameroon has allegedly been small, compared with the effect of external price shocks for instance, it has been concentrated in the relatively better-off non-agricultural and urban sectors (Subramanian 1996). The research reported here was conducted at a time of rising unemployment and reduced public services, changes that particularly affected the urban population in Douala.[4]

The women interviewed in Mabanda agreed that poverty was widespread in their community and they gave examples of a number of poverty indicators. These included an inability to meet household food needs so that members had to exist on one meal a day, not being able to afford to send children to school, basic and overcrowded housing, an inability to meet contributions to a credit society or mutual assurance group, not being able to afford hospital treatment, a husband likely to be unemployed (and possibly regarded as a financial burden) and women being forced to work in the informal sector. This poverty has resulted in a number of strategies being adopted by women.

Women's survival strategies

A critical survival strategy adopted by most of the women who were interviewed was working in the informal sector. Of the fifty-two women covered in the survey, only two had formal sector employment (both were teachers) compared with thirteen out of the fifty men surveyed. Thirty women were engaged in informal sector activities including ten who sold goods directly from their home. Of the twenty-four men considered to be working informally, only five were operating directly from the home. The informal sector, therefore, is a dominant source of income and activities relating to daily consumption items are overwhelmingly important. Twenty-seven of the thirty women and nine of the twenty-four men involved in the informal sector were engaged in the preparation or trade of food items, firewood, cigarettes or drink. Most common among these activities were the buying and selling of very small quantities of fresh or dry goods, running small bars, buying goods to process and sell (such as fish or *beignets*), growing food for sale (usually fruit and tomatoes), and prostitution.[5]

The informal sector activities of the women interviewed are typically labour-intensive and capital saving since labour is usually their principal resource. Thus, inputs and outputs are both exchanged within the local economy (that is within walking distance of the home as women cannot afford to travel outside the area), few women employ additional workers, they have little working capital and buy goods for resale in small quantities (often daily). Their children usually accompany them or are left with neighbours or relatives. Where women operate from home (possibly because of the presence of small children), the location of the house influences their income potential.

Changes in economic circumstances have also made letting out rooms an important source of income for more than 60 per cent of households.[6] Many households are obliged to let due to a lack of alternative sources of income. This has resulted in overcrowded living conditions, which were viewed by the women interviewed as an indicator of poverty. Overcrowding is also caused by members of the extended family moving in when they can no longer afford to meet rental payments elsewhere. Where this additional household member is unemployed, the poverty experienced by the household is compounded by the need to meet increased food expenditure.

Personal networks are an important survival strategy adopted by low-income women. For some their ability to cope in crisis situations when their needs exceed resources may depend largely on their 'acceptance' within the community. Hence, those who have lived for long periods in one community, have relatives living close by or belong to the same ethnic group as most of the other residents are more likely to belong to mutual assurance groups or to find people willing to assist them by looking after children or offering loans. In extreme cases when these urban safety-nets were inadequate, women return to their village or parental home, which underscores the importance of networks as a survival strategy.

Access to, and membership of, informal finance groups is an additional and critical survival strategy adopted by low-income women in Mabanda. It is estimated that up to 90 per cent of women in southern and western Cameroon are members of at least one savings group (Schrieder and Cuevas 1992). There are two main types of groups, associations and tontines.

Associations

An association is formed by a group of people (usually less than a hundred) who belong to the same ethnic, political or religious group or else live in the same area. In Douala, ethnicity and area of origin appear to be the most important basis for membership and this is revealed by the fact that some of the women who were interviewed travelled to other parts of the city in order to attend association meetings of a particular (usually minority) ethnic group. Implicitly, a further factor which members have in common is their minimum income levels which enable them to undertake identical contributions. Meetings are held regularly and a contribution (generally annual) is made to a fund which is then distributed to members who are ill or have had a death in the family. This provides an important safety-net against unexpected crisis for vulnerable urban households who would otherwise face significant extra hardship.

Associations are sometimes formed to meet specific expenses such as water or electricity supplies in the neighbourhood. Different associations may provide different 'cover' and the size and timing of contributions may also vary. At the time of the survey, most associations in Mabanda had an annual subscription of US$ 10–20.

Tontines

Credit associations or tontines are usually formed within an existing association so that in addition to an annual subscription women also contribute a fixed amount on a regular basis which they then receive in rotation. For example, a group of sixty members each contributing US$ 4 per week permits each woman to borrow approximately US$ 240 every fourteen months. Some tontines allow variable contributions providing that these add up to a fixed amount over a given period of time. As with associations, the timing and size of contributions varies between groups. While most tontines handle small amounts of money (with weekly contributions from as little as US$ 0.80 per week) the more affluent Cameroonians participate in tontines involving thousands of US dollars. Some sources suggest that tontines might handle between one-half and two-thirds of total household savings in Cameroon (Schrieder and Cuevas 1992). Although both men and women participate in tontines, they usually do so in separate groups.

While some groups save for a specific purpose such as paying school fees, others are formed to finance any individual activities. Some women said that tontine disbursements allowed them to repay loans from merchants who had advanced them goods for resale or to buy larger quantities of goods for future trading activities. Occasionally, a hybrid of associations/tontines may emerge involving a smaller annual payment, regular savings contributions and an additional contribution in the event of the sudden death within a member's family or illness of a member. Contributions may also be made to a loan fund, from which members may borrow at relatively high rates of interest (often in excess of 100 per cent per annum).

Among the survey group such finance was not used to obtain access to housing for a variety of reasons. First, the majority of respondents were already home-owners having bought land and built or acquired housing when they arrived in the area during the period of the oil boom. Second, most of the households interviewed had few needs to further consolidate their housing. Indeed, all but three of the households surveyed occupied houses made of durable materials or wood. Any further accumulation of property was beyond the means of the group at the time of the survey.

For most women, these savings groups offer the only affordable access to finance for investment or large expenditure. The women interviewed consistently stressed the benefits that membership of informal finance groups conferred on them. They described those who could not afford to contribute to such a group as constituting the poorest, most vulnerable people in the community: an ability to participate in a tontine was therefore regarded as an indicator of household poverty.

The prevalence and sustainability of rotating savings and finance associations in Cameroon, combined with poor access of women to formal financial services, has occasionally led to their use as a channel for external formal financing. For example, the German development programme, GTZ, has extended a loan to a group of twenty-seven savings groups in the Northwest Province which has enabled these groups to borrow up to three times their savings at low interest rates. Funds are, in turn, lent to individual group members among whom peer pressure assures repayment. Such schemes can work where they build on genuine established groups who can accurately judge members' borrowing and repayment capacities and among whom there is strong peer pressure. Where the funds have not been contributed by members, peer pressure can be weak so it is important to link one member's default with penalties for the group as a whole.

THE SURVIVAL STRATEGIES OF LOW-INCOME WOMEN IN ZAMBIA

In 1993, 42 per cent of Zambia's total population of 8.9 million people lived in urban areas (World Bank 1995). This high rate of urbanization is attributable to a number of factors including the importance of the mining sector and migration to the 'Copper Belt', weak agricultural policies and state intervention which resulted in a large urban-based civil service. Until the early 1970s, Zambia's economy, based primarily on copper exports, was relatively prosperous but it became increasingly unstable as international copper prices declined. The concurrent rise in the price of oil had, by the end of the 1980s, resulted in negative economic growth rates. Starting in the mid-1970s, a number of structural adjustment measures were taken to realign the economy, the effects of which were felt particularly acutely in urban areas. Consumer prices increased dramatically as a result of devaluation and the removal of price controls and subsidies. In addition, there was a substantial decline in formal sector employment and real wages, which by 1990 stood at 25 per cent of their 1975 level (O'Reilly and Gordon 1995).

As in Cameroon, the result of these declines in the formal economy was an increase in informal sector employment. Data from the Priority Survey in 1991 classified 35 per cent of women's urban employment as being in the self-employed category as compared to 18 per cent for men, and 50 per cent of urban women were unemployed compared to 25 per cent of men (Republic of Zambia 1993). Informal sector activities were widely acknowledged to be the main contributor to household earnings, particularly for the poorest sectors of the population, including women.

When questioned, Zambian women named the following as indicators of poverty: eating less frequently or going hungry; having to buy food in small (and hence more expensive) quantities; lack of meat and non-essential foods in the diet (such as tea and rice); the lack of other essentials like clothing, health, education or rent; the lack of safety-nets for crisis needs; a general feeling of powerlessness and lack of means to earn a living. The vast majority of the women had little hesitation in classifying themselves as being poor, although not necessarily the poorest even though the sample was drawn from the participants in the CARE Food for Work project which involved low status work (manual, dirty, and strenuous, with payment in kind) which would probably only appeal to those with no easier alternatives.

Women's survival strategies

Zambian women used a number of often overlapping strategies to cope with their poverty. These included: social strategies such as marriage and fertility decisions, and personal networks; general economic strategies such as migration, education and skill acquisition, links with rural areas, investment, saving and borrowing strategies and urban agriculture; and income-generation strategies which vary in the outlay and risk entailed, and hence accessibility to the poorest women. This categorization may be somewhat artificial in attempting to differentiate between activities which directly earn money and those which reduce risks, create or conserve resources for current consumption or investment for the future (Grown and Sebstad 1989). But, for the women interviewed, these activities are all components of a complex survival strategy.

As in Cameroon, the income-generation strategies of low-income women in urban Zambia are heavily concentrated in the informal sector and mostly in single-person enterprises. Women do not generally co-operate in business although sometimes a husband and wife may run a joint enterprise. Few women employ additional workers and women are usually assisted by unpaid family labour. Those who can afford to hire labour usually limit themselves to just one or two workers. The most common activity of the poorest women is selling foodstuffs, a low-risk activity which although poorly remunerative requires little working capital and generally no additional investment. Activities which require greater investment of skills and capital include the illicit distillation of spirits, basket-making, fish trading, embroidery, tailoring and running a restaurant. Some women who have access to more capital may be able to run dry goods stores, trade with rural areas, sell second-hand clothes or open a hairdressing salon.

Of course the women with the lowest incomes have no surplus to invest or save as all available income is spent on essential items of daily consumption and even this expenditure may be reduced when money is very scarce. Any small surplus that exists may be spent on consumption items for resale (which can also be used by the family if needed) or other small-scale business. While some households may own bikes or radios, few can afford anything more substantial. Similarly, for those who do not already own their homes, house purchase appears to be well beyond their means.

The research found that few compound residents used the formal banking system. Instead, women overwhelmingly rely on informal finance particularly through groups dealing with informal finance known as chilimba.

Chilimba

Chilimba groups are similar to the tontines found in Cameroon and other traditional savings schemes throughout Africa. Chilimbas consist of between four and twenty members who are usually neighbours, friends or fellow market vendors. Although chilimba groups usually comprise women, men may also form groups and among those interviewed, the largest contributions were made by groups of men. Each member makes a contribution of an agreed fixed amount at regular intervals and takes turns to access the pooled resources. Cycle lengths vary from ten days to four months. Market vendor groups tend to have more than ten members who make small, daily contributions of around

US\$ 1, a system that is well suited to the needs of small businesses with a regular turnover. An interesting variation of the chilimba groups are those which are based on contributions of work and/or meals. For example, participants in the CARE project had adapted chilimba to their 'income' source whereby two or three women donated their fortnightly food ration to each other in rotation, enabling the recipient to sell part of it to generate cash income (Henry *et al.* 1991 report the same phenomenon in Cameroon).

Chilimba proceeds are used for either business or domestic use such as the purchase of food (O'Reilly 1996). However, as in Cameroon, chilimba disbursements are rarely used to buy a house or undertake housing improvements again for various reasons. First, many Zambian respondents were already home-owners at the time of the survey, having acquired their property during a period of more favourable economic conditions, and most houses were permanent structures of brick or earth. Second, for those who are not able to own and are forced to rent accommodation, chilimba finance may occasionally be used to pay rent but, as in Cameroon, this is unlikely to occur regularly as the cycle of access to funds rarely matches the schedule of rent payments. The fact that there is a wide range of rental accommodation available means that most tenants can find somewhere to rent, although what they get may not be particularly good value for money. Chilimba is unlikely to contribute to house purchase because this was beyond the means of those interviewed in the groups at the time of the survey.

Although the chilimba is flexible, the poorest women without regular income or any surplus over their immediate consumption needs are excluded (O'Reilly 1996). However, in times of great need, money may be borrowed, without interest, from friends or relatives, albeit that in the current economic climate few people have the necessary surplus resources available to lend. Faced with few alternatives, compound residents do resort to *kaloba* or informal moneylenders who charge interest rates of 100 per cent per month and seize household assets if payment is late. Despite its harshness and the women's attempts to avoid it (for fear of being unable to repay), *kaloba* borrowing is useful as a last resort, and is readily available without formal application or collateral requirements. In general, women appreciated the discipline of informal finance groups which foster the habit of saving as well as enabling them to use this income to make special purchases.

CONCLUSION

The NRI studies in Cameroon and Zambia suggest that women's survival strategies incorporate whatever 'resources' they have at their disposal. Personal networks are used to provide childcare, access to produce from rural areas or emergency loans. Moreover, participation in informal finance groups is regarded as an important survival strategy which not only complements personal networks but also allows women to invest in informal sector activities on which they are disproportionately dependent.

Among the groups surveyed, the finance derived from informal groups is largely invested in income-generating activities. There is no explicit contribution to housing access except in the general but important sense of helping women juggle resources at times of unforeseen financial crises. In both examples, the groups surveyed were too poor to accumulate additional (or maintain existing) property resources. For example, in Zambia, where more households were renting, even rental payments would be sacrificed to meet immediate food needs during times of particular hardship. Interestingly, among the owner-occupiers surveyed in both Zambia and Cameroon, there were no indications of property being used as collateral for formal finance to fund, for instance, enterprise development. The reason for the apparent neglect of such a 'bankable' resource merits further research into issues such as legality of tenure and the weakness of non-collateral savings/lending schemes which make no use of collateral even where it is available.

Given the tight financial regime under which low-income women operate in Zambia and Cameroon, resources have to be used flexibly if unexpected and unforeseen expenditures are to be met. Informal finance groups offer considerable flexibility: women can join those groups where they can afford to make the contributions, access to funds does not require collateral or form-filling, and funds can be used for either consumption or investment. Moreover, informal groups can be adapted to the needs of low-income women as they offer access to both savings and 'insurance cover' for crises.

However, the flexibility of informal finance schemes is not boundless. The research found that those women who were unable to participate in these groups often constitute the most deprived members of the community (O'Reilly 1996). While there may be scope for more sensitive, needs-driven, participatory development of such informal groups, therefore, it is likely that the poverty faced by the poorest women is intractable in the short to medium term. Moreover, by permitting savings to be used for consumption, however necessary, such savings schemes do not protect members from a decline into deeper poverty. Savings groups are a necessary but insufficient means of avoiding total poverty in times of severe economic decline. This underscores the need, if savings and loan schemes are to be sustainable in the medium term, to place more emphasis on the funding of investment activities rather than on consumption.

NOTES

1 One could argue that the importance of the informal sector is generally underestimated due to the lack of data on two important activities: crime and prostitution.
2 Research was funded by the Overseas Development Administration (now the Department for International Development) but the views expressed here are the author's own and should not be attributed to the DFID. The author would like to

acknowledge Caroline O'Reilly, Gisele Yitamben and Alan Marter for their invaluable contribution to this research. For a more detailed account of the Zambian research, see O'Reilly and Gordon 1995.

3 At the time of the research, the last census held in Cameroon was in 1987. The economic recession which has occurred since has probably resulted in slower urban population growth with some people returning to their villages, making it difficult to estimate the current urban population.

4 Shortly after fieldwork in Cameroon was completed, the CFA Franc was devalued by 50 per cent, which must have caused further hardship among the urban population, which tends to be more dependent on imports (such as rice and wheat flour) than its rural counterpart.

5 Although none of the women interviewed described themselves as prostitutes, prostitution was mentioned in interviews as an important source of income for women. However, with economic decline, even prostitutes have had to supplement their incomes by engaging in street vending and other informal sector activities.

6 The survey focused on the main household occupying a particular house, and did not, therefore, include people renting a room/part of a house.

13

A GENDERED PERSPECTIVE ON FORMAL AND INFORMAL HOUSING FINANCE IN BOTSWANA

Kavita Datta
University of Wales, Swansea

INTRODUCTION

There is a widespread consensus among academics and development practi-tioners that extending housing finance to low-income groups is crucial if there is to be a resolution to the housing crisis in developing countries (Merrett and Russell 1994, Okpala 1994, van Huyck 1987). There is also agreement that the type and system of finance adopted (formal or informal) shapes the domi-nant form of urbanization (Renaud 1987b). This choice determines the availability of housing finance to two critical constituencies: low-income groups in general, and low-income women in particular.

Women's interests in housing have not always been recognized – an ironic fact given the patriarchal association of women with the 'home'. This marginalization of women in housing research is a cause for concern for a number of reasons. First, the persistence of female-headed households in devel-oping countries means that there are significant numbers of women who may have to acquire urban housing themselves (Buvinic *et al.* 1983, Dwyer and Bruce 1988).[1] Second, the traditional link between gender and poverty sug-gests that it is harder for women heads to acquire housing than it is for male heads (Robertson 1992). Third, gendered power relations within the house-hold can undermine the position of married women living in nuclear households. For example, given the inherently dynamic and fluid nature of households, the vesting of property rights solely in male hands makes the position of married women precarious in cases of divorce or widowhood.

In the context of the housing finance debate, the patriarchal nature of most societies and economies means that women's access to and use of housing finance is affected by their gender. This is particularly the case in formal finance systems. Low-income households are generally marginalized by stringent lending criteria such as formal sector employment, possession of legal collateral, lengthy application procedures and the tendency of institutions to extend large loans. One could further argue that women, especially those who are heads of households, are even more marginalized due to their general under-representation in formal sector employment and the fact that they rarely own or have difficulty proving ownership of urban assets which can be used as collateral. Given the double and even triple role that women perform in many developing countries, lengthy time-consuming application procedures can deter women more than men. As Berger (1989) argues, the opportunity cost of the time women spend on finance arrangements involves not only lost income (as in the case of men) but also a displacement from household labour. Moreover, specific gender-based discrimination such as requiring husbands/male partners to co-sign loans extended to women further marginalizes them.

It is estimated that formal finance accounts for a mere 5–20 per cent of housing finance in developing countries, much of which is consumed by the top 20 per cent of the income spectrum, and results in the production of no more than 10 per cent of annual national housing totals (Okpala 1994). The fact that formal finance systems are pre-disposed towards certain types of housing (such as completed, owner-occupied dwellings rather than incrementally built or rental accommodation) restricts the supply of housing which is particularly suited to low-income men and women (Merrett and Russell 1994, Osondu and Middleton 1994).

Seventy to 80 per cent of housing finance in developing countries still originates from non-institutional or informal channels (Okpala 1994). The existence of these informal mechanisms cheapens the cost of housing for low-income households and makes it more accessible to them (Osondu and Middleton 1994).

The inability of formal finance to filter down to low-income men and women alike has led researchers and policy-makers to experiment with more innovative finance programmes such as rotating savings and credit organizations (ROSCAs) which are often initiated or sustained by the NGO sector (Mitlin 1997). These innovative finance schemes are especially important as they emphasize the critical role that women play in creating, managing and accessing this resource. Indeed, some researchers suggest that there is a critical link between the extension of this finance to women and their empowerment. Women, it is argued, gain both confidence through their participation in micro-finance groups and respect from their communities for their role in accessing scarce resources (Ackerly 1995).

As this chapter argues in the context of Botswana, however, despite the apparent success of informal finance in reaching low-income men and women, as well as its vital contribution to housing production, this sector has received limited assistance and the link between formal and informal finance mechanisms remains poorly developed. Most of Botswana's financial resources are directed towards the provision of formal finance solutions to the housing crisis with particular consequences for low-income men and, especially, women. After briefly tracing the evolution of the housing finance crisis in the country, the first section of this chapter seeks to identify the reasons behind this formal approach. Second, the chapter examines why, in the face of the marginalization of low-income men and women's housing finance interests, viable and innovative NGO initiatives have not evolved. I argue that while this is partly due to the small size of the sector, it is also related to relations between the NGOs, the government and external funders. Third, and in this context, the chapter focuses upon the informal strategies adopted by as many as two-thirds of the low-income men and women interviewed to finance the construction of their homes.

The information presented in this chapter is derived from two periods of fieldwork in Gaborone in 1992 and 1996. The former involved interviews with 86 owner households living in low-income settlements, of which 34 were headed by women and 52 were nuclear. Extended interviews were held with a further 33 households, 19 of which were headed by women while the remainder were nuclear households (Datta 1994). The latter period of fieldwork was devoted to interviews with formal finance institutions, government ministries and the NGO sector. These interviews sought to investigate the provision of formal and innovative housing finance in the country.

THE HOUSING FINANCE CRISIS IN BOTSWANA AND FORMAL SECTOR RESPONSES

It may seem contradictory to suggest that a housing *finance* problem exists in Botswana, a country which has successfully transformed its economic and social landscape since independence in 1966. Classified then as one of the poorest countries in the world, Botswana was reclassified as a middle lower-income country by the mid-1980s. Unlike most developing countries, between 1965 and 1989, Botswana's GNP per head grew at an average annual rate of 8.5 per cent, foreign exchange reserves stood at a healthy US$ 3.3 billion in 1991 and government cash balances were equivalent to eighteen months of total expenditure (Harvey 1992, Ministry of Local Government, Lands and Housing 1997).[2] But, despite this rapid rate of economic growth, finance (or the lack of it) is now seen as the main obstacle preventing the smooth functioning of the housing market. In order to understand why, it is necessary to briefly trace the development of the urban housing market over the last three decades.

Starting from very low levels, the urban population of Botswana has grown very rapidly and between 1965 and 1980, the country's average annual rate of urban growth was the second highest in the world at 12.4 per cent per annum. By 1996 the urban population accounted for 48.7 per cent of the total population, 49 per cent of urban households were headed by women and a significant proportion of this population was poor (Central Statistics Office 1991, Ministry of Local Government, Lands and Housing 1997).[3] Both the rate and nature of urban growth caught the government totally unprepared as its own production of housing (through the Gaborone Town Council) was limited to white-collar public-sector workers living in nuclear households.

The subsequent establishment and rapid growth of a squatter settlement, Old Naledi, on the outskirts of Gaborone convinced the government of the need to transform itself from a provider of housing to a facilitator (Ministry of Local Government and Lands 1978). Consequently, three housing institutions were established: a para-statal organization, the Botswana Housing Corporation (BHC) in 1970; a public organization, the Self-Help Housing Agency (SHHA) in 1976; and the Botswana Building Society (BBS) in 1977. The BHC was charged with the responsibility of providing housing for both sale and rent to all income groups while the SHHA was to cater for the housing needs of low-income households through the provision of serviced land on which applicants would construct their own dwellings.[4] The BBS was to deal specifically with the housing finance requirements of middle- to high-income urban households.

Having created these organizations, and despite the rhetoric of being an 'enabler' and 'facilitator', the government proved to be unwilling to relinquish its hold on urban housing markets. Increasingly, government intervention undermined the financial viability of the BHC, SHHA and BBS which performed below expected targets. In 1991, the BHC's waiting-list consisted of 29,000 households of which 55 per cent were in Gaborone while the SHHA's backlog was close to 37,000 households (Ministry of Finance and Development Planning 1991, Ministry of Local Government and Lands 1992).

The main reason identified for this shortfall was the shortage of serviced land which the government sought to address through its ambitious Accelerated Land Servicing Programme (ALSP). This was to release a total of 33,000 serviced urban plots between 1990 and 1995. The programme ran with a reasonable degree of efficiency and with minimal corruption although it fell below its target and provided only 22,228 plots by 1997 (Ministry of Local Government, Lands and Housing 1997). The shortfall in land provision, however, was not the major problem facing the Land Servicing Programme. The ALSP coincided with the first significant attempt by the government to reduce its subsidization of urban living. Thus, ALSP applicants were required to purchase land which had hitherto been granted virtually free or at highly subsidized rates.[5]

The Botswana experience, therefore, highlights the need to provide finance for the acquisition of land, a factor that is often overlooked in housing finance research. In the context of urban Botswana, the lack of this finance presented (especially) low-income households with a major financial obstacle which was reflected in the response to the availability of ALSP plots. By July 1997, only

583 residential offers had been accepted (and deeds registered) out of a total of 13,904 offers. This financial predicament was intensified by somewhat over-zealous requirements such as the insistence on front-end payment for servicing costs, prohibitive development covenants, plot prices which were not related to affordability assessments and the short period allocated to plot purchase (Ministry of Local Government, Lands and Housing 1997).

Moreover, apart from pressuring households to acquire finance to purchase land, the ALSP also largely left plot purchasers to finance the cost of the structure or house. While the ALSP did originally earmark 51 per cent of its total budget to housing finance, it did not specify how these funds should be utilized. By mid-1989, of the US$ 64 million allocated to housing, over US$ 20 million had already been committed to goods and services unrelated to the financing of house construction. Consequently, people who applied for land were encouraged to look for housing finance elsewhere and the housing deficit which had previously been attributable to a lack of serviced land was increasingly due to a dearth of housing finance (Bank of Botswana 1989). According to Bank of Botswana estimates, between 1990 and 1997, a total of US$ 525–656 million was needed to cater for housing finance demands while the country's entire financial sector had just over US$ 315 million in resources.

Formal sector responses

The government has responded to the 'new' housing problem by focusing overwhelmingly on formal sector initiatives and continued intervention in the housing market. It is estimated that the formal sector continues to provide as much as two-thirds of total mortgage finance through organizations such as the BBS, BHC and SHHA which collectively provide as much as 92 per cent of all housing loans (Ministry of Local Government, Lands and Housing 1997). The close link between these formal institutions and the government means that these organizations have been used as instruments of government policy and have lacked organizational flexibility. Eighty per cent of the BBS's present mortgage portfolio is for housing loans and it extends 20 per cent of total national housing loans (or 53 per cent of total lending). The average size of loans is US$ 5,249 which is repayable over a maximum period of twenty-five years. The BHC, a *de facto* supplier of housing finance since the late 1970s, has lent approximately US$ 3.4 million, approved over 500 housing loans and sold over 2,000 houses at a total sale value of US$ 9.2 million.[6] It provides 5 per cent of all loans which account for 18 per cent of the total lending, at an average loan size of US$ 6,756. A major player in the housing market, the SHHA has facilitated the production of 27,000 houses, invested over US$ 2.6 million in urban infrastructure and provides 67 per cent of all housing loans but only 2 per cent of the total amount lent.

The critical point to be stressed is that the government has favoured formal sector initiatives in the face of the housing finance crisis. Led by the Botswana Democratic Party, the government has cast itself very much in the role of the benevolent state which takes 'care' of its citizens, particularly the more articulate and vocal urban residents. In the past, this role has translated into initiatives such as the subsidization of BHC rents and BBS mortgages and the provision of land through SHHA at virtually no cost to applicants. Moreover, all three organizations are themselves heavily dependent upon 'soft' government loans issued through the Public Debt Service Fund. The financial cost of this 'benevolence' has slowly become apparent as urban growth has accelerated and the government has been caught in a predicament of its own making. Opposition parties have been quick to capitalize on this and have organized popular protests against any rent or rate increases and encouraged people to default on their payments. The fact that the economy is doing well has further fostered a sense of dependence, as people increasingly demand that the government continue in its self-appointed role as provider. Despite government rhetoric on the spirit of self-help in the country and the long-established tradition of (rural) people providing housing for themselves, the urban population has been unwilling to give up any of the privileges that it has traditionally enjoyed.

Such a large government commitment to long-term funding is increasingly seen as problematic in government circles given that it ties up resources for long periods of time and diverts money from other social investments. Moreover, this approach makes it very difficult for other housing finance organizations to emerge unless they are given the same preferential treatment. The government has attempted to reduce its involvement in housing finance, but in such a way that favours formal initiatives through the private sector. These institutions have been far from innovative in their approach to housing finance. Just as the financial capacity in Botswana is limited, so is the physical capacity to deliver finance.[7] There are few financial organizations due to a relatively unsophisticated financial system and, in comparison to housing, alternative investments such as short-term trading have been highly profitable and have captured a great deal of commercial bank finance (which had an excess liquidity of US$ 52.5 million in 1989). Conversely, the attractions of investing in the housing sector are weak given extensive government interference, high and increasing costs of construction throughout the 1980s and a common perception that urban housing is over-priced. Government intervention has slowed down the flow of savings to housing finance and diverted them to alternative higher-yielding investments. As Okpala (1994) argues elsewhere, this has particular consequences for low-income households as with less saving there is less lending and with less lending it is not usually low-income households who get access to loans. It is to this that we now turn.

Impact of formal finance on low-income
men and women

Formal finance has had a minimal impact on urban low-income men and women and there is a perception that the government favours and funds organizations whose loan portfolios are biased towards richer urban households. The BHC and BBS, in particular, have exhibited this bias partly due to distrust of the ability of low-income households to honour their loans. The general opinion is that low-income groups have had little exposure to paying back and managing mortgages and one respondent at the BBS went so far as to suggest that a culture of repayment did not exist among low-income men and women. To be fair, these opinions are based on the experience of the BBS and other formal finance institutions during the construction boom of the 1980s when lending criteria were relaxed.[8] With the end of the boom, the BBS was left with clients who could not repay their loans and foreclosures did increase, confirming the perception that low-income borrowers were not reliable managers of money.

Default has also emerged as a significant problem in the SHHA, whose housing programme is directly targeted at low-income groups. Current estimates put about 65 per cent of recipients of SHHA housing finance in arrears. Although the average amount of money owed is small, such a high rate of default does not inspire confidence in formal finance institutions and encourage them to go 'down-market'. For example, the BBS has been lukewarm in its response to recent initiatives to extend smaller home improvement loans to lower-income households living in SHHA areas.[9] Interviews with formal finance institutions revealed a common perception that extending finance to low-income households was best left to the government and NGOs who had access to 'cheap' finance. The BBS in particular has cultivated an image of being a finance organization for the rich to the extent that most low-income urban households do not even attempt to apply for loans. This is borne out by the fact that the society's lending has been limited, due as much to a lack of qualified clients as to the volume of work or lack of available funds.

While it is marginalizing low-income households in general, the tendency of the government to rely on formal-sector based solutions to the housing finance crisis is especially harmful to women. This is due to gender discrimination both in employment and the housing market. This argument is backed by other studies which have found that although 'gender per se is not a good predictor of access to financial and other forms of assistance it does covary with poverty and related factors which inhibit access to, and effectiveness of assistance' (Mogwe 1992). Various studies in Botswana illustrate the relative poverty of female-headed households when compared to male-headed households (Bond 1979, Koussoudji and Mueller 1983). The formal sector (with the exception of health and education) tends to be dominated by men in Botswana who, on average, also earn higher salaries than their women counterparts.

Mogwe (1992) estimates that there are only 200 women for 2,100 men who earn over US$ 262 per month and that urban male-headed households have 2.6 times the earning power of urban female-headed households. Moreover, as women start work at lower wages and their earnings peak earlier than men, the gap between male and female earnings tends to widen with age (Brown 1980).

The majority of women interviewed in 1992 were employed in the informal sector as street vendors and domestic servants, a sector in which earnings are both erratic and low. Unemployment levels are also higher among women with 41.7 per cent of urban women unemployed as compared to 20.7 per cent of men. The higher wages earned by men and the preponderance of women among the ranks of the unemployed reveal the persistence of an ideology which views men as breadwinners despite the high incidence of female-headed households in the country. Thus, formal institutions that use income-based requirements such as formal employment, or a regular minimum income four times in excess of rent/mortgage repayments, and the large size of the average loan, render many women ineligible for their assistance.

Gender discrimination in the housing market is best illustrated in the case of married women living in nuclear households. Under both customary and general law, husbands have 'marital power' over their wives, which means that the BBS, BHC and commercial banks all require married women's spouses to co-sign their mortgages. Female-headed households also suffer gender discrimination in the housing market. For example, the BHC's stipulation that only the earnings of a husband and wife constitute 'household' incomes discriminates against female heads who may either be the sole earners in their households or where the second worker is a child/sibling. In either case, they are effectively excluded from BHC schemes.

The only formal finance organization which appears not to have marginalized women is the SHHA. In total, one-third of the households surveyed in 1992 had acquired housing finance from SHHA with 33 per cent living in nuclear households and 35 per cent in female-headed households. However, although these figures suggest that a relatively high proportion of women heads have acquired finance from SHHA, a further examination of this population found that many of the recipients were the older women in the sample whose husbands had been alive when they built their dwellings. The fact that, generally, so few households bothered to obtain housing finance from the SHHA is attributable to a number of factors including the lengthy waiting period and the fact that loans were generally inadequate to construct a house given escalating building material prices, high standards of construction and the construction boom of the 1980s. This led to the emergence of a significant finance gap and 25 per cent of nuclear households and 21 per cent of female-headed households interviewed in 1992 who had acquired loans had to supplement them from elsewhere.

The distribution of formal finance reveals a considerable bias against both low-income men and women. This raises the question of how these households have coped with either the total or inadequate provision of housing finance by the formal institutions.

INNOVATIVE MICRO-FINANCE?

In many developing countries, the finance gap that has emerged out of the inadequate provision of formal housing finance for low-income households has been filled by innovative housing finance schemes often involving the NGO sector. This has significant gender implications given that women are often given a central role in such schemes resulting both in a greater access to finance as well as, on some occasions, their empowerment. Unfortunately, in the case of Botswana, the NGO sector has played a relatively limited role in the provision of urban housing finance. Indeed, the idea of a partnership between the government, private sector and established NGOs has not even been discussed (Ministry of Local Government, Lands and Housing 1997). This can be attributed to two reasons.

First, in its role as provider, the government has always maintained a somewhat suspicious attitude towards the NGO sector. This can be discerned from its attitude to women's NGOs such as Emang Basadi. This NGO, which deals with issues such as the political empowerment of women, has been accused of trying to overthrow the government and form a Woman's Party. The strained relationship between the two sectors is further revealed by the very limited government funding of the NGO sector. Again, this can be largely attributed to a government belief that it is best placed to deliver assistance to the people even though NGOs have the potential of reaching low-income women. The lack of internal funding has meant that the local NGOs have been very dependent upon external funding (from international NGOs and donors) which in turn has had an impact on the viability of their programmes. As Botswana is no longer classified as a low-income country, many donors are now in the process of winding up their operations, leaving local NGOs with the unenviable task of finding new funding sources. Almost all the NGOs discussed below have had to deal with this crisis in the availability of external funding. Possible partnerships with the private sector have been identified by some local women's NGOs but they are aware that they have to deal with discrimination against their interest in 'women's issues' which are often seen as being unduly confrontational, divisive and in some extreme cases unnecessary.

Second, the slow and recent growth of the NGO sector has meant that there have been few urban-based NGOs with whom the government can engage in a dialogue. It is estimated that between 1985 and 1989 the number of local NGOs in Botswana rose by 60 per cent but the sector only became organized around the time of the Beijing Conference when an umbrella organization, the National NGOs' Coalition, was formed. Within this loose coalition, NGOs which have been specifically concerned with gender issues have formed their own registered NGO, the Women's NGO Coalition. The formation of this umbrella organization has strengthened the position of local NGOs *vis-à-vis* the government as they can now 'speak with one voice'. NGO networking has also meant that there is less duplication of projects and wastage of resources.

More recently, NGOs have begun to concentrate on urban-based initiatives, although NGO finance for housing remains limited compared to income-generation programmes. For example, the Women's Finance House (WFH) was established by a group of professional women in 1989 specifically to provide micro-finance to women. It was started and sustained by external funding from the United States, Norway, Sweden and Denmark – through USAID, the Norwegian Agency for Development Co-operation (NORAD), the Swedish International Development Agency (SIDA) and the Dutch NGO HIVOS respectively. Loans by the WFH were first issued in 1993. Within three years, the WFH had extended loans to 311 women and basic business training to 675 women entrepreneurs. Loans are extended at market rates but operate without the stringent criteria used by banks such as collateral, existing savings accounts in excess of US\$ 131, and business records. The average WFH loan size is US\$ 472 and the maximum amount that can be borrowed at any one time is US\$ 1,312. However, the rate of default on loans is high (up to 85 per cent in Gaborone) despite regular follow-up visits by WFH workers. Common reasons given for default include members selling goods on credit which cannot be recovered, the use of funds for non-productive purposes (such as crisis needs like funeral expenses) and the high cost of living in the city. The fact that group lending has not emerged as common practice among NGOs (although see below) may be identified as another possible reason for a high rate of default. As loans are extended to individuals, as opposed to groups of entrepreneurs, group pressure cannot be bought to bear upon defaulters. Recently, the WFH has extended a few house improvement loans primarily as an 'add-on' to income-generation programmes as only applicants whose businesses operate from home qualify. Housing loans are typically small and geared towards making limited improvements.

On the whole, the WFH and other NGOs have remained wary of extending housing loans (especially for construction) due to a perception that the sums of money involved are relatively large and a reluctance to engage in a long-term issue given the complications of access to land in the post-ALSP housing market. While some have not gone much beyond the talking stage (for example, the Co-operation for Research, Development and Education and the Housing Action Trust), others, like the Botswana Christian Council, have had to end their housing programme due to the withdrawal of external funding in 1993. Cancellation is especially unfortunate given that the Council built 400 houses between 1981 and 1995, had very low rates of default, and worked with particularly marginalized women. This low rate of default is attributable to the fact that the council worked very closely with its members, kept its costs down by requiring households to contribute their own labour to construction, and also involved local brigades in the production of relatively cheap building materials. Moreover, it linked its housing programmes to income-generation activities. Thus, for example, small households were encouraged to occupy one room and sub-let the other while other women were encouraged to engage in petty trading and street vending. Perhaps more significantly, rates of default among recipients may have been low due to the religious nature of the organization and the fact that many of the women in its projects attended church.

Currently, the only NGO which is extending housing finance is Habitat for Humanity (HFH), which was established in 1991. HFH's first venture was in a rural setting in Kanye although it has recently begun various urban programmes in Kasane, Ghanzi and Francistown. It has yet to extend its operations to Gaborone where the need for housing finance is greatest. The unique feature of HFH operations is that it works with communities, as opposed to individuals, a strategy which, as already indicated, other NGOs have yet to adopt. Research elsewhere has found that women's participation tends to be significantly higher in projects which use solidarity group mechanisms (Berger 1989). Groups are required to build a demonstration house (with locally raised finance) before they are considered for loans. Once the demonstration house has been built, prospective 'home-owners' are identified: usually those who have access to land and who have participated in the construction of the demonstration house. Prospective owners are also required to contribute 100 hours of labour in the form of 'sweat equity' before construction can begin on their own homes.

A revolving fund is then set up based on a US$ 13 contribution per member per month and each applicant receives between US$ 1,050 and US$ 1,312 which has to be repaid over a ten-year period. Default on loans has not emerged as a significant problem, partly because of the swift response of local committees who repossess property and exert communal pressure on defaulters through the local courts (*kgotlas*). In total, approximately 250–300 houses have been built but once again HFH is heavily reliant on external funding (from USAID in this instance) which draws into question the long-term viability of the programme.

In general, the government has preferred not to engage in discussions with the handful of NGOs involved in thinking about housing finance. Given the general lack of NGO finance for housing, therefore, we need to review the strategies that low-income households have adopted to finance house construction.

INFORMAL HOUSING FINANCE STRATEGIES

Low-income households have adopted a variety of strategies to access informal housing finance. Although treated separately here, often households use a combination of strategies which finance first access to, and later the consolidation of, housing. These strategies are gendered in that there are differences in the manner in which women heads raise finance for construction as opposed to men. The fact that most households engage in individual as opposed to community strategies can be attributed to the fact that many low-income people live in the same settlement simply because they have been allocated plots there, there is a constant movement of a tenant population in low-income settlements, and the grid layout of the settlements inhibits the creation of a sense of community. Hitherto, the dependence on the government has also inhibited the formation of informal, community-based organizations.

Savings

According to the survey data, 35 per cent of nuclear households and 24 per cent of female-headed households in Gaborone relied on savings to construct their houses. In part, the lower proportion of women heads using their savings was attributable to the fact that as men earned higher salaries, nuclear households were able to save more. Moreover, the household income of these households was also higher given that one-third of them had both the husband and spouse working. The corresponding number of female-headed households who had a second income was lower. Irrespective of household structure, however, the total savings of all households was fairly small and certainly insufficient to construct more than a very basic structure. Thus, the majority of households built incrementally when money became available. Such a building process is ideally suited to a housing market which lacks an efficient housing finance system and where access to finance is limited to only certain urban social classes (Osondu and Middleton 1994). Women heads, in particular, had built temporary structures (such as mud or zinc sheet huts) when they moved onto their plots (68 per cent) as opposed to 50 per cent of nuclear households.

Although the incremental building process ensures that low-income men and women build at a pace that they can afford, it is often a very slow process, particularly for female-headed households. This is significant in the case of urban Botswana as many of the households interviewed were building during the 1980s when construction costs and building material costs were escalating. A study conducted by the Ministry of Local Government and Lands in 1992 found that households earning below US$ 52 per month would take roughly four years to complete their homes. In my sample, a high proportion of the women earned lower wages and took much longer to consolidate their dwelling.

The incremental building process is tempered in the case of Gaborone by SHHA's requirement that temporary structures such as mud huts be demolished within two years of applicants being allocated plots, to be replaced with structures built using modern building materials. There has been some criticism of this requirement given that it exerts a pressure on households to build at a pace which they may not be able to afford. Moreover, over a period of two years, temporary structures can consume scarce resources which are lost when they are demolished. Yet, one has to recognize that a significant proportion of urban households want to live in houses built with 'modern' building materials, perhaps due to a desire to be seen as modern themselves (Larsson 1988).

The use of 'modern' building materials poses a particular problem to women heads and further raises the cost of construction. Although women have the skills to construct traditional homes, they do not necessarily possess the relevant skills to build more modern housing.[10] Consequently, the preference for and requirement of modern houses has meant that low-income households have to hire professional labourers. Again, this is gendered, with 53 per cent of female-headed households relying on professional labourers as opposed to 17 per cent of nuclear households. In addition, women heads rarely contributed their labour to the construction process which again raised the cost of construction. This can be attributed both to a lack of skills as well as of time due to their position as primary earners. A common complaint from female-headed households was that labourers took longer than necessary to construct their rooms as they were unsupervised. One woman who was interviewed said that she was sure that the labourers were working elsewhere during the day when she was not there even though she was being charged.

Although female-headed households may take longer to consolidate their dwellings and incur more costs, they do succeed in constructing the houses that they desire. This is perhaps best illustrated in the case of Boitumelo, a woman householder who was living in Gaborone at the time of the interview.

Boitumelo

Boitumelo was 38 years old at the time of the interview. She had two children who lived in her village with her mother. She came to Gaborone in 1968 and squatted on land in one of the city's low-income settlements, Old Naledi. With the initiation of the SHHA, she was allocated a plot in another settlement, Bontleng. Thus, she was able to take advantage of the earlier SHHA schemes which allocated land free of charge.

Having acquired a plot, Boitumelo built a temporary hut and then obtained a building material loan from SHHA which enabled her to start constructing the main house. However, the loan was insufficient to build more than a very basic structure. At the same time she started to brew and sell beer (initially from her plot) which brought in some money and later she set up a second vending

---

business. At the time of the interview, she was earning US$ 105–131 per month.

Gradually, Boitumelo has been able to save enough money to expand and consolidate her home. Each time she has added to her dwelling, she has had to hire labourers as she does not have the time or skills to do so herself. The last time labourers were hired (in 1991), she paid them US$ 262. She couldn't remember how much she paid for the building materials.

Although Boitumelo complained that the building process had been slow, she was satisfied that she had eventually succeeded in building a house she could be proud of.

In general, married women fare better than women heads as these households often build their dwellings themselves and will only hire labourers if they are relatively well-off and can afford to do so. Moreover, men are able to contribute labour while their wives can supervise the process which means that the hired labourers are much less likely to steal any building materials or waste valuable time. Florence's story illustrates some of these issues.

Florence

Florence was 44 years old at the time of the interview and was married. She came to Gaborone in 1974 in order to pursue a secretarial course and stayed with her uncle. In 1975 she met her husband and moved on to the plot that he had obtained from SHHA in Bontleng. Initially, it was hard to raise the money for construction as only her husband was working. So, they approached SHHA for a building material loan and built a small room which they used for sleeping, eating and washing. Her husband built the room with Florence's help as she was not working. Florence then decided to start working herself. Since she had just had a child, she decided to work from home and started sewing clothes. Gradually, her business took off and at the time of the interview she was earning US$ 131–157 a month. In addition, her husband had been able to find a job as a technician in a government-run enterprise.

As their financial situation improved, Florence and her husband decided to extend their house. They decided that they could now afford to hire labourers and as Florence was working from home, she could supervise them. By the mid-1980s, they had extended their original one-room house to a five-roomed house with a separate kitchen and bathroom/toilet.

These two case studies show how the households combined a range of strategies to finance the construction of their homes. The finance gap created by an inadequate building materials loan was supplemented in both cases by informal finance in the form of income from alternative income-generating activities. However, significantly, the relative cost of construction for the two households was different with Boitumelo having to rely much more on hired labour which Florence and her husband were able to avoid until they could afford to hire this labour.

Social networks

A second strategy adopted by households to raise finance is to rely on the help of relatives and friends, either for money to enable them to undertake further construction or in terms of labour. A higher proportion of female-headed households rely on help from their relatives or friends in raising the finance for the construction of their homes as well as help in the form of a labour contribution. Finance is usually extended in the form of a loan (generally interest free with flexible terms of repayment) or as an outright gift (if the amount involved is not exorbitant) as shown by Grace.

Grace

Grace came to Gaborone in the late 1970s and shared accommodation with her sister who was living there at that time. By the early 1980s, she had acquired a plot in the same settlement and begun the process of construction. Since she was working as a day-labourer, her income was very low so she was only able to construct a mud hut. Having worked for some time, Grace decided to start to construct more permanent rooms. She was still working as a labourer and had very limited savings as she had to support her child. So, Grace approached her sister for help and was given a loan of US$ 398. Grace supplemented this with her savings which amounted to US$ 131. This was sufficient to begin the construction of two rooms.

Having built the rooms, Grace realized that she faced an uphill task in paying her sister back. So, she decided that instead of moving into the rooms herself, she would let them out while remaining in her mud hut. At the time of the interview her rental income was approximately US$ 39 per month. Eventually, once the loan had been repaid, she hoped to be able to move into the rooms as her own accommodation was very poor.

The fact that women can rely on the help of their relatives is partly based on the traditional practice that women can expect/demand help from male relatives and their extended family in return for fulfilling family labour requirements (Glickman 1988, Kinsman 1983, Brown 1983). Although these systems are undergoing change as the basis of economic production has moved away from the family unit to the individual, one can still see the extended family operating over considerable distances. Thus, for example, young women and men coming to urban areas can rely and indeed expect their urban-based kin to house them until they have found a job which will enable them to rent a dwelling or save enough to construct one of their own (Datta 1996). The 1991 census estimated that 7 per cent of the urban population lived in such shared accommodation (Central Statistics Office 1991). Female-headed households are better placed to ask for family assistance than women living in nuclear households where there is an understanding that the husband should be able to provide for his dependants. Monyati is one of the women interviewed in Gaborone.

Monyati[11]

Monyati was 25 years old at the time of the survey and had a child who was living with Monyati's mother in her home village. She had moved to Gaborone two years prior to the interview to look for a job. On arrival in the city, she moved in with her brother who was married and had two children. She was still living with them at the time of the interview. Her brother was a vet and earned about US$ 262 per month. Monyati was still unemployed.

Monyati said that she was sharing with her brother because he was her only relative in Gaborone and as such she had to rely on him. He had been very kind to her and let her share a room with his children whom she looked after. In exchange for living in her brother's house, Monyati also does most of the domestic chores.

When asked if she would ever move into her own house, Monyati responded that given her financial situation, it was unlikely that she would ever earn enough money to begin construction. But she did think that if she got a job she would be able to move in to a rented room where she would have greater independence.

The existence of these social networks and their use by female-headed households is indicative of the relative poverty of these women who were often unemployed or lacked a regular income. In urban areas, sharing is a form of tenure with fairly permeable boundaries and interviews revealed that respondents moved in with relatives when unemployed. The motives of relatives for providing accommodation are often a recognition that they started their urban lives as sharers and that given their own tenure arrangements (many are renters) and precarious employment conditions (many work in the informal sector) they may need to move into shared accommodation themselves.

Rental income

Once a basic structure has been built, a third strategy to raise (additional) housing finance for the consolidation of what many respondents termed the 'main house' is to sub-let rooms. Over one-half of the female-headed households interviewed pursued this strategy. In some cases, women heads let the temporary dwellings which they had constructed when they first moved onto the plot, or the better rooms were rented as these fetched higher rents while they stayed on in the mud huts. Tsholofelo was one of the female-headed households interviewed who had adopted this strategy:

Tsholofelo
Tsholofelo worked as a domestic servant at the time of the interview. She had four children, two of whom were living with her mother in her village while the older two lived with her. Tsholofelo had moved onto the plot in 1970 and built first one and then a second room. When she got a job as a domestic servant (which included accommodation) she decided to let her house so that she could supplement her work income (approximately US$ 79 per month) with rental income. She wanted to invest her rental income in building a new house for herself which would have a flush toilet, separate bedrooms and living space. At the time of the interview (roughly twenty-two years after she had moved onto the plot) Tsholofelo had begun the construction of her 'main house'.

However, not all single women experience Tsholofelo's problems in constructing her home. As shown by Cecilia, some women heads are able to expand their rental business quite quickly and earn considerable amounts of money. It is important to note these variations so as to avoid falling into the trap of viewing all female-headed households as marginalized 'victims'.

Cecilia
Cecilia was 32 years old at the time of the interview and had moved onto her plot in 1985. She was a dressmaker and earned roughly US$ 210 per month. Apart from her dress-making business, Cecilia also had a street vending stall which brought in additional income. She had four children all of whom stayed with her. In addition, she had younger siblings living with her. In total, there were fourteen people living with Cecilia.

Cecilia financed the initial construction of her house by borrowing US$ 236 from her employers (she was working as a clerk then) and supplemented this with her own salary. She was able to build two rooms and a kitchen, and subsequently added a third room when her businesses became more profitable. By the

---

late 1980s, Cecilia decided to begin a rental 'business' and started constructing additional rooms. By the time of the interview, she was letting three rooms which were bringing in an additional US$ 92 per month. She planned to use the money to improve her own dwellings as well as support her family.

Nuclear households are less likely to let rooms than female-headed households partly because they do not experience the same economic necessity to do so as their income is generally higher. The nuclear households which do rent rooms are often those where the wife is not working and therefore can manage the rental business (Datta 1995). Some of these points are illustrated in the case of Belinda.

Belinda

Belinda was 58 years old at the time of the interview. She worked as a nurse while her husband was unemployed and had been for some time due to ill-health. She has three grown up children who live with her together with their children. In total, there are eleven people in her family.

Belinda and her husband did not initially want to rent accommodation but in the early 1970s they were approached by an old friend who was in desperate need of accommodation. They decided to allow her to construct a temporary structure at the back of the plot. When her friend left, Belinda was able to purchase the structure from her for US$ 45 (which she admits was a good deal but 'things were cheaper then') and was able to let the room.

She subsequently decided that it would be a good idea to extend their rental accommodation as the extra money could be invested in the construction of their own house but also go towards meeting their immediate needs such as food and clothing. The second room cost US$ 80 to build and the third one slightly more. At the time of the interview, Belinda thought that starting a rental business had been a good decision.

Rental income has emerged as one of the most important sources of income for low-income households in Gaborone and is often invested in housing consolidation. It is of particular importance to female-headed households who while likely to operate at a smaller scale than nuclear households, are very committed to their business and recognize the importance of the extra income in achieving the dream of a main house built of modern building materials.

CONCLUSION

The debate over the provision of housing finance in Botswana has centred almost exclusively on formal-sector initiatives, whether public or private. The government has been instrumental in leading this debate given that it currently provides much of the housing finance in the country. One can argue that much of this government intervention has been inappropriate and misdirected. In the first instance, extensive government interference has prevented the development of thriving financial systems. In such a scenario, it is almost inevitable that low-income households have little chance of acquiring formal finance. Second, government intervention has overwhelmingly assisted middle- to high-income households at the expense of their poorer urban counterparts. Formal finance institutions have largely ignored government initiatives designed to encourage them to go 'down-market'. Third, there has been little recognition of the fact that a significant number of urban households are headed by women and that these households are likely, on average, to be poorer than male-headed households. Female-headed households have had limited access to formal finance not only due to their economic situation but also, more importantly, due to their gender and household composition. Indeed, it is not only women heads who face constraints in acquiring finance but also married women who still require their husbands' consent before they can secure a loan.

Given the gender-blindness of formal finance provision, appropriate government interventions should focus instead on two initiatives. First, it could encourage the development of informal finance mechanisms. NGO programmes have proven to be a success elsewhere especially when one considers gender issues. Focusing on community participation, these programmes have achieved low default rates in spite of the fact that they work with marginalized low-income households. Moreover, the extension of finance to and through women to low-income communities, and the grassroot nature of these programmes, have meant that increasingly access to finance has come to be associated with empowerment.

In the context of Botswana, there is evidence that such informal finance mechanisms can work. Certainly, the Botswana Christian Council's housing programme was unique in that although it operated among very poor female-headed households, it had very low rates of default. However, the nascent nature of the NGO sector and the hitherto uneasy relationship between it and the government have inhibited the development of a dialogue. This could change given the greater organization of the NGO sector as well as government recognition that formal finance mechanisms are excluding a significant proportion of the urban population. The withdrawal of foreign funding might mean that such a dialogue could be accompanied by greater domestic financial support of this sector and its activities.

A second, seemingly contradictory, initiative would be to forge links between informal and formal finance mechanisms. As Okpala (1994) argues, the enormous housing deficit across the developing world means that, despite the important and significant contributions of informal finance, this system cannot cope with the task alone. Given that loans are typically small and geared towards incremental housing construction, informal finance systems are usually associated with low rates of construction. Moreover, such finance is often provided at a very high cost (economic and social) and does not link low-income women and men to mainstream financial systems (Berger 1989). Indeed, one must be careful not to romanticize the impact of informal finance mechanisms. Research is beginning to show that far from empowering marginal men and women, it may lock them further into greater debt, an increased fear of social exclusion, and higher workloads. The latter has a particular impact on women who may already have double, or even triple, workloads but are still expected to assume responsibility for servicing household loans. The forging of a link between formal and informal mechanisms would *technically* give these households the best of both worlds as well as translating into higher quality housing and infrastructure services. The challenge in Botswana is to convince both the formal and informal finance institutions of the benefits of engaging in such a partnership.

NOTES

1 Although see Varley (1996) who disputes the numerical significance of these households while emphasizing that the overwhelmingly (feminist) concentration on female-headed households (and indeed certain types of these households) renders women living in nuclear and extended households invisible.

2 US$ 1 = Pula 3.81

3 The large proportion of female-headed households in Botswana can be attributed to the historic emigration of men to South African mines which delayed both the age and rate of marriage in the country. More recently, the relative importance of marriage has declined as women have found an economic and social space within urban areas which enables them to maintain their independence (Datta 1998).

4 Low-income households are classified as those earning between US$ 472 and US$ 2,625 per annum.

5 The price of ALSP plots is fixed by cabinet and is based on either cost recovery or market costs depending on the target group. Two types of schemes operate: for low-income households the average plot price is US$ 787 while for middle low-income households (US$ 2,625–3,937) the average plot price is US$ 1,470. Plots are sold at the market price to middle- and high-income groups (CORDE 1994).

6 Finance was first provided through the Tenant Purchase Scheme which involved the sale of rental housing to existing tenants. The impact of this scheme has been somewhat limited given a common perception that BHC housing is over-priced and largely unaffordable. This is also reflected in the response to BHC finance which, although provided at very favourable rates when compared to other financial institutions, has had very few takers. Over three-quarters of the purchasers of BHC housing acquire finance from elsewhere.

7 In order to address this problem, the government has recently been involved in discussions on the possibility of establishing a housing finance institution in the country.

8 The BBS portfolio increased dramatically in 1989 when the amount of money lent in housing mortgages rose by almost 50 per cent as compared to an increase of half this level between 1980 and 1989 (Bank of Botswana 1989).

9 This is partly due to land tenure problems whereby SHHA 'plot-owners' have the right to use and develop the land but ultimate ownership belongs to the government. Although this land can be mortgaged, in practice most financial institutions will not accept it as collateral.

10 The construction of traditional housing in Botswana has long been regarded as women's work which is indicated by the fact that houses are often referred to by the names of women residing in them. However, there is a strict division of gender roles and clear demarcation of men's work and women's work, so that if a house is plastered in cement rather than clay, for example, then construction is a man's job (Enge 1982).

11 This case study is derived from a third period of fieldwork in Gaborone in 1995 which looked at the impact of gender on shared accommodation.

14

ENGENDERING FINANCE

A comparison of two micro-finance models in El Salvador

Serena Cosgrove
Northeastern University, Boston, USA

INTRODUCTION

Micro-finance for both income-generation and housing construction has received increased attention from development practitioners, governments and official banking systems since the 1970s. Internationally recognized models of micro-finance lending include the communal banking system of the Grameen Bank in Bangladesh with over US$ 1.5 billion in loans and a 97 per cent repayment rate, as well as the ACCION micro-finance network which includes BancoSol, a commercially successful bank for low-income men and women in Bolivia (Bornstein 1996).

Over time, a number of debates have evolved questioning best practice, the bench-marks of successful programmes and priorities for funding. This debate has revolved around the merits of 'finance-only' and 'finance-plus' programmes and between financial sustainability and socio-economic impact. Each side has its own proponents. While the World Bank, USAID and representatives from the official banking system argue for finance-only programmes, other more progressive agencies such as Oxfam (UK-Ireland), Oxfam America and Bread for the World advocate programmes that facilitate micro-finance growth *and* community development. The World Microcredit Summit held in February 1997 promised to provide a forum in which these issues could be further discussed by participants who wanted to emphasize the importance of micro-finance as a development tool. However, summit organizers chose to concentrate upon minimalist finance-only models, thus producing a simplification of the best approaches and models for micro-finance lending.

The limits to the micro-finance debate are problematic for a number of reasons. First, while the finance-only approach is attractive to conservative donors given that the sustainability of the programme loan portfolio is guaranteed by the number of borrowers served, it concentrates on providing finance and little attention is given to the structural, political, economic or cultural factors which promote poverty. Models which promote the finance-plus approach, while less efficient initially, promote social and human development in addition to economic change in the long term. Second, the emphasis on the finance-only approach focuses our attention away from the diversity in lessons and triumphs of other micro-finance lending. This is a dangerous path as one model cannot respond to the locally determined needs of informal entrepreneurs, cultural idiosyncrasies or the historical, economic and political factors that affect solidarity and finance usage. For example, people living in countries recently emerging from dictatorships or post-war situations may need more than finance-only programmes to facilitate the growth of micro-enterprises or housing starts. The same can be said for countries undergoing severe economic problems. Finance-only programmes will not aid borrowers' ability to address the structural reasons for their poverty and their harsh living conditions. Instead, while such programmes help micro-enterprises to survive from one day to the next, additional services such as technical assistance, training and follow-up which are necessary for their long-term development are neglected. Third, and perhaps most significantly, gender issues have to be considered when deciding on appropriate programmes given that a significant proportion of micro-finance borrowers are women. The subordinate status that women occupy in many societies means that they may face obstacles towards which finance-only programmes (which can be gender-blind) are not sensitive. Finally, the decision to simplify the complexity of micro-finance programmes makes it difficult for alternative models to access international donor funds.

This chapter argues that there is no best practice of working with micro-finance. Instead, local conditions determine the kinds of finance and/or services that clients need. It examines two models of micro-finance in El Salvador, which allows the author to refocus the international debate in Salvadoran terms. Here, the nascent nature of democracy, the increased poverty and the trauma of a long and bloody civil war in which 80,000 Salvadorans were killed and many more injured mean that development alternatives have to take a different social, political and economic reality into account. The two models of micro-enterprise lending primarily serve women and were developed by two Salvadoran NGOs, Fundación Salvadorna de Apoyo Integral (FUSAI) and Corporación de Proyectos Comunales de El Salvador (PROCOMES). Of the models presented here, the FUSAI approach is the closest to the one recommended by the World Microcredit Summit while PROCOMES is based on a finance-plus approach.

The fieldwork for this chapter comprised qualitative and quantitative methods including interviews with various stake-holders such as NGO staff and management, and donors, as well as with borrowers and non-borrowers living in the same communities. The impact of the two programmes on women micro-entrepreneurs was assessed by examining the experience of thirty borrowers with four communal banks in San Salvador. The stories of four of these women will be presented as case studies.

The chapter illustrates that although finance-only and finance-plus approaches have their own merits, the more comprehensive finance programmes facilitate long-term consciousness-raising and empowerment. Hence, the latter have a more significant impact on women borrowers, their families and communities. This conclusion facilitates a more textured reflection on the best approaches for micro-finance because whether one is in Indonesia, Bangladesh or the United States, effective services to micro-entrepreneurs often necessitate more than just a small loan.

MICRO-FINANCE AND GENDER: MEASURING EMPOWERMENT

Prior to developing the micro-finance debate further in the context of El Salvador, it is necessary to detail the key connections between micro-finance, gender and empowerment given that the vast majority of borrowers are women and micro-finance is increasingly being seen as a tool of development.

Past research has suggested that women are somehow empowered by their participation in finance schemes even though the term 'empowerment' is one of those wily words which is used by different practitioners to mean different things. A report produced by the International Research and Training Institute for the Advancement of Women (INSTRAW) reminds us that the term 'empower' is derived from the word power:

> Power is defined as the ability . . . to act or perform. Empowerment is defined as the capacity to exercise control. Economic empowerment should be defined as the individual or collective capacity to solve matters pertaining to work and its cumulative value.
>
> (INSTRAW 1995: 6)

Development strategies targeting women have not always been over-concerned with questions of empowerment. Projects inspired by the 'women in development' (WID) movement advocate economic reform without any systematic or structural change to the process of development itself. Inspired by liberal feminist discourse, proponents of WID argue that the subordination of women within the household can be redressed if women are given the opportunity to undertake paid work. Thus, women become empowered as they perform paid work to meet both their households' and their own immediate and more long-term needs. Furthermore, as women dedicate proportionally more of their income to the household than men, an increased income for women means better education, nutrition and housing for all household members but especially for children (Blumberg 1995).

However, I would argue against conflating increased income with the empowerment of women as it remains unclear whether the women control this new income. Moreover, the paid work that they undertake leads to an extension of their working day as they continue to engage in domestic work. Indeed, one could argue that women remain marginalized in household, community and municipal decision-making as even though they may experience increased income individually, they do not gain access, control or use of community resources. The weakness of the WID approach, therefore, is that it avoids the integration of women into the development process in terms of project planning, implementation and evaluation.

The alternative approach, known as 'gender and development' (GAD) moves away from the 'add women and stir' scenario implicitly advocated by the supporters of WID and towards a different appreciation of empowerment (Young 1993). Proponents of GAD call for the transformation of socially constructed power relations between men and women. In essence, this approach argues that if women are to achieve equality with men and become active agents in the home and wider political and development arena, they will need more than just income-generating projects. Moreover, GAD projects do not single out women but propose that through the development process both women *and* men can learn new ways of sharing power that is more equitable. Consequently, GAD projects typically entail consciousness-raising with both women and men as integral participants in this process.

GAD proponents argue that while income may be a significant indicator of empowerment, it is clearly not the only one. Even if a programme increases the income of clients, this does not necessarily imply that they are less vulnerable (economically and socially) or that they have been able to renegotiate unequal gender relations. Moreover, questions such as whether the poorest men and women in a community are incorporated in a project or whether women entrepreneurs increase their access to and use of resources go unanswered. By contrast, therefore, the GAD approach deals with a set of more difficult, but also more significant, indicators such as changing gender relations at the household level, decreases in domestic violence and the increased participation of women in community affairs.

Understanding the differences between WID and GAD is essential to the ways we measure empowerment and the effects of micro-finance. Adopting a WID approach would encourage the implementation of minimalist finance-only schemes on the premise that an increase in income is the determining factor in women's empowerment (Blumberg 1995). Indeed, one suspects that the popularity of the WID approach is the relative ease that it offers to measuring programme impacts by collecting data on income-based indicators such as loan recovery rates or increases in the number of borrowers served.[1]

MICRO-ENTERPRISE LENDING IN EL SALVADOR

Throughout Central America, Salvadorans are recognized for their entrepreneurial spirit. From exclusive shopping malls to the informal markets of downtown San Salvador, one gets the impression that the whole of El Salvador is a large bustling market. For informal vendors, however, this vibrancy is not due to economic growth but rather to intensifying levels of poverty that are forcing increasing numbers of people to work on the streets, in the markets and in the home. In addition to the impact of the civil war of 1980–92 on the country's infrastructure and living conditions, structural adjustment policies and a commitment to a neo-liberal strategy of 'modernization via privatization' have translated into decreased government spending on important social programmes such as food subsidies and other services. Consequently, former public servants who once had job security and middle-class lifestyles are now joining the ranks of the un- and underemployed seeking survival through informal strategies.

It is estimated that currently over 60 per cent of the economically active population of El Salvador works in informal income-generating activities. As elsewhere in Latin America, women comprise the majority of income earners in this sector and most of them are single heads of households (Zamora 1993, Ortíz 1994). Left in charge while their men either fought in the civil war or fled from the fighting, women have had to juggle the triple load of domestic, community and income-generating responsibilities. Moreover, they have had to do this in a social, cultural and political environment which creates more obstacles than opportunities. For example, women living in low-income urban communities often face extensive sexual harassment and petty crime as they carry out their income-generating activities. However, given the important role women play in the economy and the well-being of the Salvadoran family, both government agencies and NGOs are now targeting finance services to them.

Government cutbacks have made NGOs particularly active in various micro-finance programmes. There are currently over 500 NGOs operating throughout the country of which about fifty specialize in providing financial services to urban and rural low-income people for income-generating activities. For some of these NGOs, finance is not only a means of providing funds to people marginalized by the official banking system, but also a tool to facilitate consciousness-raising, provide training opportunities and organize the disenfranchised. The shared history of struggle and hardship during the war also motivates NGOs to promote 'consciousness-raising' (*promover concientización*)[2] in the communities in which they work. Given that over 75 per cent of micro-finance borrowers are women, many NGOs working in this sector have had to acknowledge that gender is an important factor in the problems of marginalization. For many NGOs, this implies a process of training which enables women and their communities to implement their own projects, raise funds and inform the government of their needs.

Although there are many different models for acquiring finance, in most cases micro-entrepreneurs form communal banks or solidarity groups with community members or other micro-entrepreneurs who work in proximity to them. This is done with the support of finance promoters employed by the government, NGOs or private-sector finance associations. The concept of 'solidarity' is important as it allows extremely low-income borrowers who do not have any collateral to rely on other members of the group to guarantee their loans. Typically, once communal banks are formed each micro-entrepreneur receives an initial loan ranging from US$ 50 to US$ 100 with repayment periods of four to six months. Some schemes include obligatory savings programmes of 5–10 per cent of the amount loaned. Although NGOs charge higher annual rates than formal banks (roughly 36 per cent annually for small loans as opposed to a bank charge of 21 per cent), they still fill an important finance gap for micro-entrepreneurs as loan-sharks charge up to 10–20 per cent interest per day while the formal banking system continues to discriminate against potential low-income contributors.

Loans are commonly used for direct investment in the borrowers' businesses for things like repairs, raising levels of inventory or the purchase/hire of machinery. Research also shows an important link between the availability of micro-finance and housing improvements especially if the house is used as the location for an informal business. Thus, increased income generated from loans is often invested in home improvements in the form of repairs, additions and service connections like electricity and water. The borrower's ability to undertake these housing investments has a positive impact on his/her self-esteem and family well-being. The two NGOs that I will be concentrating on here are PROCOMES and FUSAI. Although both are involved in micro-finance, they have adopted radically different approaches.

PROCOMES

The Corporación de Proyectos Communales de El Salvador (PROCOMES) was founded at the height of the civil war in 1988 when its primary mission was to meet the basic needs of people fleeing the fighting in the countryside for the relative safety of the cities. With the signing of the peace accords in 1992, PROCOMES realized that it now faced 'new and very demanding challenges'. Consequently it embarked upon a series of social and economic development projects such as access to micro-finance, mutual help housing in which community members worked together to construct their own houses, environmental projects ranging from reforestation to garbage disposal programmes, and the provision of childcare and health clinics.

PROCOMES now has an integrated vision of development in which access to finance goes along with the participation of communities, the empowerment of women and an organized community which sees itself as an active agent fighting for change. Thus, the three transversal elements of PRO-COMES work that inform their programme planning, implementation and evaluation are popular education, gender and community organization. Gender is seen as being a particularly critical factor given that 90 per cent of communal banks are composed of women and 75 per cent of PROCOMES beneficiaries are women. Given PROCOMES' commitment to client-led empowerment and gender issues, they have designed a pilot project, MASA, which integrates gender concerns and finance.

The two communal banks of MASA are different from other PROCOMES banks in that gender informs all training, follow-up and decisions made on loan amounts. The banks take the gendered reality of women micro-entrepreneurs into account in the following ways. First, the project tries to avoid burdening women with too many, and poorly timed, meetings by agreeing upon the frequency and timing with all members. Second, loan amounts and training topics are also determined by members in conjunction with promoters. Third, the project has a family integration component whereby clients and their families can participate in recreational activities that promote the transformation of unequal gender relations at the household level.

Finance is organized through communal banks for the lowest income micro-entrepreneurs as well as solidarity groups made up of more prosperous micro-entrepreneurs. Extremely low-income households initially access small loans through the communal banks and then subsequently obtain larger loans through solidarity groups and in association with other micro-entrepreneurs within the same field. Loans are extended for micro- and small businesses, housing and the acquisition of appropriate technology. About 50 per cent of associated micro- and small business loans go to women and 28 per cent of housing loans go to women.

PROCOMES delegates the responsibility of selecting new members for the communal banks to local finance committees formed by volunteers who are also micro-entrepreneurs. In the case of MASA, the local finance committee of the municipality of Nejapa recommended eight women. When the bank was formed, the members attended four pre-finance meetings that touched upon such subjects as the history of PROCOMES, the nature of its work, the alternative financial system and a participative diagnosis of the gendered nature of life for each member. The subordination of women is seen as being a particularly important issue in these meetings given that it can inhibit the democratic participation of women in communities. In addition to home and business visits, this exercise allowed the MASA team to determine the finance and social needs of the members.

As a whole, PROCOMES has some way to go before it achieves financial sustainability within the next few years. Although MASA has a 100 per cent loan recovery rate, the other PROCOMES communal banks have a 75–80 per cent recovery rate. By 1997, PROCOMES had a small but diverse loan portfolio of US$ 300,000 and served 2,000 clients. The following two case studies seek to illustrate the extent to which PROCOMES has succeeded in its finance-plus approach to micro-finance. The two women interviewed were MASA clients and had received loan disbursements ranging from US$ 50 to US$ 115 from PROCOMES. I accompanied the PROCOMES promoter to Nejapa twice a week as she visited each member while they conducted their income-generating activities. The objective of these visits was to provide more extensive follow-up than in the other PROCOMES communal banks.

Ana Gertrudes García

I met Ana Gertrudes on my visits to Nejapa, the site of the MASA project. An energetic woman, she was sure that the acquisition of a loan from PROCOMES meant that she could go back to work. Formerly, she had worked as an ambulatory street vendor, taking buses around the country and selling cosmetics door to door. Within five years, she had developed a wide network of contacts and meeting the demand for her merchandise was a challenge. She stopped working when she had children.

At the time of the survey, she had decided to go back to work as her children were now older. However, as she was renting a house in Nejapa with her husband, Cristobal, Ana Gertrudes could not save enough money to make the initial investments to build up her inventory. This is when she approached PROCOMES who agreed to give her a loan of US$ 100 through the MASA communal bank pilot project. Having received the loan and in consultation with members in her church, Ana Gertrudes decided that instead of going back to long days on the road, she would invest in the purchase of cosmetics and become a mini-distributor. She started distributing goods to her friends in her church and collecting money from them once they had sold what she had given them. In conversation with me, she commented: 'I'm glad that I do not have to travel so much, but I'm particularly happy about the increase in income and the independence.'

Since receiving the loan, Ana Gertrudes has diversified her business to include the sale of vitamins. She has increased her income, much of which she has spent on her children. She has also begun to make plans to buy a house. Before she obtained the loan, she did not think that she could afford monthly house payments. Now, she thinks this may be possible. Nonetheless, her increased income has created friction with her husband. As soon as he saw that she had more money, he used it as an excuse not to give money for household shopping. Consequently, although she enjoys her increased income, it is now hidden from her husband so that he continues to give her money.

Esperanza Centeno

When she moved in with her boyfriend, José, 21-year-old Esperanza happily retired from working in a *maquila* assembly plant producing shirts for export. After working for three years 'under constant pressure', she was glad to quit her job and to move into the two-room house that José had built for them. Economically dependent upon José, she stopped giving her parents money and began to spend her days preparing food, cleaning the house and washing their clothes.

When Esperanza and José heard about the micro-finance programme of PRO-COMES, they decided that Esperanza should take a loan in order to start some paid work to generate more household income. José came up with the idea of selling firewood although Esperanza had no prior experience in running a micro-enterprise. Esperanza was accepted by the MASA communal bank and was given US$ 100 to get her business off the ground. Soon after receiving the loan, however, the road in front of their house was dug up for the installation of a sewage system and the firewood was never delivered. Moreover, Esperanza found that she was pregnant. Consequently, José loaned the money to a neighbour who needed some quick capital and took over making the payments on Esperanza's loan.

As the first loan repayment period came to an end, José and Esperanza began to talk about whether or not she should re-apply for the next loan cycle. José said that they could now buy the firewood because the delivery truck could make it to the house. Esperanza told him that she would like to raise chickens in addition to selling firewood and they fenced the back yard in preparation for this. Esperanza plans to send money to her parents from the sale of her chickens and to increase her participation in the household economic decision-making process. Although she could not use her first loan, Esperanza did participate in all the meetings and scheduled training sessions. Esperanza feels that participating in the training has helped her to express her needs. She commented, 'the training makes me feel like something is waking up inside of my mind that I did not know before.'

These two case studies reveal how participants of finance-plus schemes may begin to experience empowerment. This is perhaps best illustrated in Esperanza's case: one can see how over time, she has become more independent. She now has more of a say in household decisions (as opposed to the start when José seemed to make most of the decisions) as well as some control over the way in which her income will be spent (she intends to start sending money to her parents again). Ana Gertrudes' case illustrates the development of friction between her husband and herself as she has started to earn an income. This is a good example of how increased income does not immediately lead to empowerment as Ana Gertrudes' husband expects her to use her money for household necessities while he spends his income on himself. Ana Gertrudes has obviously developed a strategy whereby she hides her true income from her husband thus forcing him to continue to contribute to the household budget. This is partly the reason why PROCOMES has decided it is important to involve clients' spouses and male relatives in organized activities so that they can reflect on their behaviour and maybe begin to change.

FUSAI

FUSAI was also founded during the civil war (in 1987) with extensive funding from the United Nations High Commission for Refugees (UNHCR). FUSAI's mandate was to work with refugees who had been either internally displaced or forced to flee the country to refugee camps in Honduras during the Salvadoran army campaigns in the countryside. Once the peace accords were signed in 1992, FUSAI responded to the emergency conditions of the returning refugees with donations, temporary shelter, food and finance. However, it subsequently decided to move from working only with the refugee population to working with low-income groups in urban areas, principally through the extension of micro-finance.

FUSAI now has a communal banking programme which provides micro-enterprise, housing improvement and construction loans. Micro-enterprise loans are disbursed to either individuals or *colectivos* (communal banks). The latter consist of groups of six to ten members usually made up of friends, relatives or neighbours. When disbursing through *colectivos*, FUSAI charges an interest rate of 3.5 per cent per month on the amount earned. FUSAI does not prescribe to the philosophy that the group 'co-signs' for all its members but rather it requires co-signers from outside the *colectivo* or collateral in the form of household appliances or land/house deeds. As *colectivo* promoters repeatedly told me, they are not interested in facilitating deeper levels of solidarity among members. In a forty-five minute *colectivo* orientation before loan disbursal, a FUSAI promoter said:

> At FUSAI we don't talk about solidarity like in other credit programmes. It does not exist, it's not real. People take advantage of the security blanket of solidarity; it is seldom that someone stands up to defend the interests of some Christian as in the case of the good Samaritan.

Rather, FUSAI uses established bonds of kinship or friendship to guarantee loans: a loan co-signed by one's mother-in-law is unlikely to experience default.

Although over 90 per cent of the *colectivos* members are women (58 per cent of clients are women), and 40 per cent of the individual loans and 67 per cent of the housing loans are extended to women, FUSAI limits its actions to providing finance in the most timely and efficient manner possible. It leaves social problems, consciousness-raising and organizing around community rights to other institutions.

Initially, FUSAI found working with finance frustrating due to difficulties in maintaining acceptable loan repayment rates.[3] Currently, the *colectivos* have an 85–90 per cent loan recovery rate while rates are higher for individual and housing loans (87 and 95 per cent respectively). Due to an apparently efficient administration and accounting system, FUSAI is able to cover all costs of the finance programme as well as 40 per cent of the institution's costs through the interest charged. In 1996, FUSAI had a total loan portfolio in excess of US$ 1.7 million with 5,000 clients.

The two case studies below illustrate the experience of two women who have received FUSAI loans.

Francisca Ochoa

Francisca, 54 years old, is a natural leader and a micro-entrepreneur. She sells milk products at an open-air stall near the Apopa municipal market. I met her for the first time when she brought Maria Magdalena, José Carlos and Rosa Elba to attend a *colectivo* orientation meeting. She and the three others were interested in forming a *colectivo* and receiving low-interest loans through FUSAI. My initial impression of Francisca was her strength. She did not hesitate to ask the FUSAI promoter questions when something was not clear. Often women do not ask men questions, especially when the man is seen as being from a 'higher' class. It was clear that the four of them had many responsibilities and they did not want to dedicate any additional time beyond taking care of the formalities. Consequently, they were not interested in training as far as I could tell. They just wanted small loans in a quick and efficient manner.

Francisca had been financing her micro-enterprise with high-interest loans from one of the many moneylenders working in the Apopa market. Paying an exorbitant interest rate of 10 per cent per day (compared to 3.5 per cent per month with FUSAI). Francisca felt that she was 'dándole de comer al gordo' (feeding the fat man) instead of feeding her own kids. She accurately conveyed the consensus of the group: 'We want a break; we want out from under the loan-shark.'

A little over a week later, the loans were ready for disbursement. Gathered at the FUSAI office in Apopa, José Carlos, Maria Magdalena, Rosa Elba and Francisca prepared to sign the contracts and take their cheques to the local bank to be cashed. Francisca turned to me and told me that she was getting a headache – she was unable to read and write and could barely write her initials, which does not qualify as a signature in Salvadoran banks. Holding her head in her hands, she said to me, 'This is my father's fault. He used to say that it made no sense to send a girl to school just so that she can learn to write letters to her boyfriend . . . it just hurts my head that I cannot sign my own name.' Francisca had to ask one of the other members of her communal bank to accompany her to sign the cheque.

My next meeting with Francisca was when I was accompanying the FUSAI promoter on his business visits as he filled out the forms for the loan. Outside of an already bustling market building, sellers set up boxes and baskets occupying the streets in their allotted areas under black pieces of plastic strung up to protect vendors and shoppers from sun and rain. First I walked past Rosa Elba and her vegetable stand, then past Maria Magdalena and José Carlos with their vegetables before I saw Francisca, waving a fly-swatter over a metal box filled with farmers' cheese and little plastic bags of ricotta. Francisca gets up at 5 a.m.

---

and catches a bus to the outlying municipality of Aguilares where she walks out into the countryside to buy the cheese. By 9 a.m. she is at the Apopa market where she stays until everything is sold, when she usually returns home (about mid-afternoon).

As I followed her progress after the loan disbursement, she told me that sales were up and that she had increased her inventory. 'I feel rested working with FUSAI', explaining how she felt less economic stress now that she was not working with the loan-shark. Before she 'was drowning working with the loan-shark'. Francisca's success is all of her own making as she gets very little support from her abusive husband with whom she has hardly spoken in six years – he eats and sleeps at the house but that is it. Francisca confessed that she is tired of having to support a man who does not earn his own living.

Juli Abigail Iglesias

Juli, a 49-year-old evangelical woman who sells shoes from a stall within the Apopa market, suggested the name for the *colectivo* she and four other women formed: 'Love of God'. The *colectivo* is made up of Juli, her daughter, Soñia, daughter-in-law, Teresa and two friends, Eva and Ana. Soñia, Teresa, and Eva have adjoining stalls where they sell dishes, clothes, and sewing materials. Juli has a stall in a different part of the market and sells clothes. Like Francisca, Juli is a strong woman who works hard to contribute to her household's well-being. In fact, she is the primary income earner in her household because her husband has not been able to work due to health problems. Soñia and Teresa (and their households) live with Juli and her husband near the highway between San Salvador and Apopa.

Initially, Soñia, Teresa, and Juli asked for loans that they planned to invest in increasing their inventory. However, just four days after their loans were disbursed, one of Juli's son-in-laws, Joaquin, was run over and killed by a bus as he crossed the highway near their house. The driver of the bus abandoned the vehicle which was subsequently confiscated by the police. Anxious to get his bus back in circulation, the owner of the bus offered Juli's daughter, Maritza, US$ 350 in compensation for the loss of her husband. Unemployed and widowed with three children, Maritza qualified for legal aid. However, the lawyer assigned by the state had eighty other similar ongoing cases and she told Maritza to get out of her office when she stopped in to see how the case was going. According to the recently widowed Maritza, the lawyer told her, 'You are only interested in the money, not in justice.'

Juli gave half of her loan to her bereaved daughter for the funeral expenses. In addition, she went to the FUSAI office to see if they knew of anyone who could help her find another lawyer. They told her they could not help. Since then, Juli has had difficulty in making her loan repayments because of the increased costs due to the accident. In a conversation with Juli and Maritza at their house, they said that one person a week was run over on the highway outside their house. Musing out loud, Maritza said: 'Well, if one person a week is killed at the curve near our house, how many people a week are killed if you add up all the curves on the road from San Salvador to Apopa?'

> Maritza finally tracked down a human rights lawyer who mediated a settlement between Maritza and the owner of the bus. He agreed to give her US$ 1,000 over a period of ten months and she agreed to drop charges against him. The case still continues against the driver of the bus, but it could be years before it gets to trial. In the meantime, Juli has had great difficulty in making her loan payments. Sales have been low because of the time of year and her frequent visits to government offices accompanying Maritza. Juli knows that the possibility of low-interest loans, like the ones granted by FUSAI, are key to the success of her business, but she and her family live so close to the border between survival and 'going under' that she was not able to take full advantage of the loan.

As I watched, listened and participated with Francisca and Juli in their income-generating activities, I was left with the clear impression that they needed much more than just finance (even though comprehensive finance programmes are expensive and often require clients to dedicate scarce free time to training which both Francisca and Juli are unable to spare). Juli's experience with the FUSAI loan, for example, is interesting because her story speaks to the necessity of addressing social problems that women micro-entrepreneurs face in urban El Salvador. Francisca needs a literacy programme and would be an ideal candidate for leadership training. She also wants to know more about her rights and options regarding her abusive spouse. Gender relations, particularly within Francisca's household, are also strained as she has taken over the role of primary provider while her husband has contributed little towards household maintenance.

CONCLUSION

The experience of PROCOMES and FUSAI illustrates that there are no easy answers and no *one* best approach to micro-finance. In fact, both approaches have strengths and weaknesses. In the case of El Salvador, finance-only loans are monitored efficiently by FUSAI which has developed a strong accounting and administration system while PROCOMES, in addition to extending micro-finance, is also committed to understanding and addressing the subordination of women in their work. Interestingly, although the micro-finance debate suggests that finance-only programmes can achieve financial sustainability more easily (by streamlining their training period and issuing as many loans as possible), the MASA pilot project not only has the highest loan recovery rate of any of the programmes considered in this chapter but it contributes to the financial sustainability of households and communities, and to the sustainability of PROCOMES' loan portfolio. This implies that finance-plus programmes such as those implemented by PROCOMES are not necessarily less sustainable as suggested by the general debate on approaches to micro-finance.

Moreover, the emphasis of finance-plus models on empowerment is partic-ularly pertinent given that a high proportion of micro-finance borrowers are women. Finance-only programmes which focus solely on income-generation cannot guarantee empowerment as they do not challenge gender roles and power relations within the household. Women's vulnerability within the household and the community can only be addressed through the process of consciousness-raising exercises with which finance-plus programmes engage. Thus, finance-plus programmes can ultimately be more cost effective because they plant the seeds for long-term empowerment of women, men, their fami-lies and communities.

NOTES

1 Of course, even monitoring income is not as easy as it may seem. As Johnson and Rogaly (1997: 73) argue: 'respondents may give false information as loans have been used for a purpose other than the stipulated one; establishing a causal relationship to the actual loan involves a knowledge of all the beneficiary's sources and uses of funds; and it may be difficult to establish what would have happened if the loan had not been made.'
2 For example, in the case of women-entrepreneurs, consciousness-raising can translate into NGO workers holding discussions with women after watching a video on domes-tic violence.
3 FUSAI workers believe that militant communal bank members offered membership to people in return for political loyalty and told them that they did not have to repay their loans.

15

HOUSING FINANCE
AND EMPOWERMENT IN
SOUTH AFRICA

Joel Bolnick and Diana Mitlin
People's Dialogue on Land and Shelter, South Africa and International
Institute for Environment and Development, UK

INTRODUCTION

This chapter describes the strategies and achievements of a South African people's organization whose membership consists predominantly of women. The South African Homeless People's Federation is a network of housing savings schemes that began in 1991 from an informal exchange and information programme between squatter areas. It is supported in its work by the People's Dialogue on Land and Shelter, a South African NGO. After some years of small-scale localized savings within the schemes, the People's Dialogue obtained capital for a housing loan programme and set up a dedicated housing fund known as the uTshani Fund. Lending for housing construction through group-based finance was started in April 1995 and over 700 housing loans have been distributed.

The Homeless People's Federation operates in a context in which most other institutions are concerned with formal sector government-funded solutions to housing needs. By contrast, the Federation has sought to demonstrate an alternative people-centred strategy for development which, while placing a specific focus on housing, also creates and supports a political movement able to represent those who previously lacked representation. The housing finance activities of the South African Homeless People's Federation, the People's Dialogue on Land and Shelter and the uTshani Fund, therefore, need to be analysed from a social development perspective. The underlying philosophy behind such initiatives is that finance provides a structure for poverty reduction and community empowerment that is effective because it simultaneously enables 'members of the Federation (to) practice self-reliance . . . (to) run their own organizations with their own resources . . . an affirmation of the dignity and strength of the homeless poor' (People's Dialogue 1996: 13). In working to support a people's empowerment agenda, the People's Dialogue and the Federation have been drawn to housing finance because of its potential to consolidate the organizational and financial skills required for community control of development. Such an approach is in sharp contrast to those for whom the provision of finance is no more than a mechanism that enables the poor to increase their incomes through better integration in commercial and financial markets.

As discussed in Chapter 2, the contribution of housing finance to social development is four-fold. First, investment contributes directly to the improvement of housing and living conditions by reducing the amounts spent on repairs and the effectiveness with which households allocate resources. Second, finance programmes are able to combine training and technical support for income-generation activities, both directly to reduce lack of income and to ensure that housing improvements are made more affordable. Third, group saving activities strengthen the informal networks within low-income settlements. These networks are especially important to women who have traditionally been responsible for improving and maintaining the home and providing basic services. Finally, financial systems and skills are essential for communities to challenge professional development agencies.

This chapter explores the extent to which these four perspectives have been met by the South African Homeless People's Federation. The chapter begins by outlining the policies and programmes of the South African government with respect to low-income housing before summarizing the strategies of South African NGOs within the fields of housing development and housing finance. The chapter then provides a short overview of the history of the People's Dialogue and the emergence of the South African Homeless People's Federation. This is followed by a detailed discussion of a number of key themes, namely: housing improvement; local economic development; community networks and women's participation; and interaction with external agencies (that is, professional agencies based outside low-income settlements, first at a national level and then in a local context). A final section summarises the experiences to emerge from the work of People's Dialogue and the Homeless People's Federation.

LOW-INCOME HOUSING IN URBAN SOUTH AFRICA

There are now an estimated 15 million people squatting in shacks in towns and cities in South Africa, and huge inequalities in housing provision across the country. For example, in 1993, it was estimated that the median floor area per person in South Africa was 11 square metres per person; there was an average of 33 square metres for each white person compared to 4–5 square metres for the black population living in squatter settlements (Mayo 1993). In part, these figures reflect the long history of housing support for better-off groups and the much more limited assistance provided to those on low incomes (People's Dialogue 1993). Recent government support for housing improvements has been heavily influenced by political events, notably the end of the apartheid government and the democratic elections in April 1994.

There have been some national and local government and NGO initiatives to provide low-income housing, but all have been small-scale and restricted to a few settlements. At the beginning of the 1980s, for example, government finance was made available to the better-off black and coloured urban residents, with some 175,000 households accepting loans before rising interest rates and civil unrest resulted in large numbers of loans being in arrears or in default (HIC/ACHR 1994). For those with lower incomes, the government funded a sites-and-service programme that was implemented by the Independent Development Trust, but only 65,000 sites had been prepared by 1994. Following these disappointments, a range of initiatives were considered to enable black urban dwellers to obtain housing loans. The National Association for Civic Organizations (SANCO), for example, negotiated with the commercial banks and the Association of Mortgage Lenders, and the Urban Foundation (then an NGO) initiated a Home Loan Guarantee Fund to enable commercial lenders to offer smaller loans. The National Housing Commission (now replaced) offered loans at market interest rates for low-income households developing serviced sites, and several para-statal agencies offered subsidized finance. More recently, in post-apartheid South Africa, the number of such programmes and projects has increased although it remains far below the scale of demand and, with a few exceptions, most are inappropriate to the needs of low-income households.

The drafting of the Reconstruction and Development Programme recognized that housing had to be a priority for South Africa. In October 1994, some six months after the election, a multi-stakeholder conference resulted in the Botshabelo Accord, which created a framework for 1 million houses to be built each year. As a result of this accord, the government agreed to offer financial and institutional capacity to private contractors and the banks who, in exchange for access to a capital subsidy programme, would be required to create housing opportunities for the homeless and make formal finance available to those who could afford it. Lobbying by the Federation and the People's Dialogue resulted in a government pledge to support what they called 'the People's Housing Process', but in the months that followed the government failed to demonstrate any significant support for this approach. Thus, the commercial contractors have sought to obtain the maximum subsidy of R15,000 (for housing, infrastructure and services, with an additional R2,500 for site difficulties) but provide only a 20 square metre house of poor construction quality. Even so, in the financial year 1995–6, the Housing Department spent only 16 per cent of its budget with the remainder being returned to the Treasury and in 1996 only an estimated 30,000 houses had been built, with 139,000 under construction and 56,000 at the planning stage (Mthembi-Mahanyele 1996).[1]

A major problem faced by the subsidy programme has been the difficulty of distributing funds. While the department put in place two measures to support housing development for lower-middle income households (targeted particularly at black and coloured professionals), the Building Warranty Scheme to control housing quality and the Mortgage Indemnity Fund to indemnify the banks, little was done to address the housing needs of the poorest groups.

For low-income communities, the capital subsidy system has proved problematic in several respects. First, private-sector contractors have sought to monopolize finance. For example, when the Eastern Cape provincial government sought to modify the subsidy regulations in order to offer subsidies to the Federation, private sector representatives condemned the Department of Housing and tried to reverse the amendment through the stakeholder consultation group, the National Housing Board. Private-sector control has also meant that relatively little by way of housing, infrastructure and services is provided for the subsidy finance. Faced with limited alternatives, however, many squatters are willing to sign away their subsidy in return for only minimal improvements. In Riversdale, a small town in Cape Province, the council had contracted a developer to build approximately 250 houses. Local residents believed that if they did not sign the contracts they would have to vacate their plots. As the residents pointed out, they knew what the contract involved and that the houses would be 21 square metres, but they did not know how big this was, nor were they offered any alternative (People's Dialogue 1997).

The second problem for low-income communities is the bureaucracy and regulations associated with the programme. In the case of one housing co-operative, such was the quantity of documentation required to access the subsidy that a photocopier was taken to the site in order to speed up the process. Elsewhere, settlements have been denied subsidies unless they can prove legal land tenure. But, many informal settlements are not being regularized as the Department of Land Affairs is not working in co-operation with the Department of Housing, so many households are being excluded from subsidy provision.

By 1996, there was a growing awareness within government that the subsidy programme was insufficient to address housing need. Consequently, the government established the National Housing Finance Corporation (NHFC) to provide housing development loans to a range of private and not-for-profit companies. However, from the start of this initiative it has been evident that the main focus is on providing loans on commercial terms to institutions able to serve those who can address their housing needs through conventional market-based solutions. The NHFC does not provide support to those who require additional finance for housing to be affordable, nor is it able to support the development of institutional capacity to assist the 15 million South Africans who are not being reached by present programmes.

It is within this context that the People's Dialogue and the South African Homeless People's Federation have sought to provide an alternative housing option. They have not been alone. Many other groups working in the areas of housing, urban development and low-income finance remain concerned about the inability of government policies to address the needs of such groups. The approach adopted by the People's Dialogue, however, is probably unique in placing primary emphasis on community-based learning and support for autonomous saving institutions linked through exchange programmes. The next section summarizes the activities of South African NGOs within this sector before examining the work of the People's Dialogue in more detail.

NGOs, HOUSING FINANCE AND URBAN DEVELOPMENT

South African NGOs have long been interested in working in urban development issues, mostly with a focus on protecting the urban poor. NGOs such as the Surplus People's Project in Cape Town attempted to defend the rights of black households to seek work and settle within urban areas through supporting local resistance and offering legal advice (HIC Women and Shelter Network 1991). However, within the confines of apartheid politics there were few opportunities for NGOs to create more proactive and progressive development solutions to address the needs of the black and coloured populations living in urban areas.

During the early 1990s, these conditions changed and the availability of capital subsidies for urban upgrading (mostly through the Independent Development Trust) involved NGOs in facilitating a process of urban development and providing advice to community leaders (Abbott 1996). In response to this demand for their services, a number of NGOs began to expand their technical and professional role to assist low-income communities to access government funding. In 1992, for example, one of the largest South African NGOs based in Johannesburg was working in twenty township upgrading projects and a further three upgrading projects in informal settlements (Planact 1992).

Some NGOs sought ways to understand better the development potential of the small-scale informal rotating savings and credit associations (called *stokvels*, *gooi gooi* and *umgalelo*) and burial societies which have long operated in low-income settlement and often involve women (Smets 1996a, Thomas 1991). At the beginning of the 1990s, Planact was actively exploring and researching into such initiatives and was also supporting one civic organization to work toward developing larger-scale savings and credit initiatives (HIC/ACHR

1994). Other South African NGOs such as the Group Credit Company were interested in drawing on international micro-finance initiatives to support the development process among groups that had traditionally been excluded from the formal financial sector.[2]

During the 1990s, NGOs working in housing and urban development went though a considerable adjustment. The civic organizations (known as 'civics') began to seek new roles through local government elections and, as has been the case in other countries in a process of democratization, key NGO staff moved across and took up government posts. Moreover, following on from the promises of the new ANC government, civics and other community associations have focused specifically on housing finance. The Group Credit Company (now established as a banking institution known as Cashbank) and the Rural Finance Facility started to provide housing loans to some of the lowest-paid workers in the formal sector who guaranteed repayments with their provident funds (Mitlin 1996).

Other long-established NGOs have focused primarily on obtaining government subsidies to support low-income housing developments. The Development Action Group (DAG) in Cape Town, for example, shifted from being primarily a volunteer agency working to resolve land struggles to being a professional agency which increasingly worked with communities on housing delivery. DAG has identified two main areas within their current work programme that are related to housing finance. First, at a project level DAG has been assisting communities that wish to access government subsidy funds. Conventional access to housing subsidies requires the developer to provide some form of bridging finance because expenditure has to be incurred before the subsidy funds can start to be drawn down. Second, DAG has assisted residents in Villiersdorp to set up a rotating housing fund to provide household loans in a mutual aid housing programme (Mitlin 1996).

While many NGOs are working under difficult circumstances and learning from their experiences, there are some concerns with the work within the housing sector, in particular, the fact that most NGOs work through the local civics and fail to represent equally all of those living in the black townships. The civics have a primary focus on the better-off residents with established rights of residence and are often dominated by men. Over 85 per cent of housing loans distributed by the Rural Finance Facility are given to men (Mitlin 1996).[3] Two groups are therefore under-represented: women (despite their household responsibilities) and squatters (among the poorest households in the townships). Second, most NGOs and civics focus on ensuring either that the government provides better housing or that the capital markets better serve the needs of those on low incomes, but they do not aim to strengthen the political capacity of the urban poor to define and articulate their own needs. It was in response to these perceived shortcomings that the People's Dialogue on Land and Shelter established its initiative in 1991.

THE PEOPLE'S DIALOGUE: A PEOPLE'S INITIATIVE

The People's Dialogue was established as a response to a strongly expressed need for a system of networking among urban informal or squatter settlements. From the very beginning there was the intention to support a people's development process whereby primary support would be given to strengthening autonomous people's organizations. At a conference of community leaders at Broederstroom in 1991 two critical issues were explored. First, did the leadership within informal settlements believe that a future democratic government would meet the development needs of those living in informal settlements? The community leaders recognized the value of local organizations in the struggle to organize to obtain housing and expressed a reluctance to wait for political liberation before housing would be delivered. Second, what did international experience in addressing the problems of access to land and housing have to offer to urban community leaders in South Africa? To answer this question a key relationship was formed with an Indian people's organization and NGO that attended the Broederstroom conference. This relationship has provided multiple opportunities for the sharing and exchange of techniques, methods, approaches and strategies (People's Dialogue *et al.* 1994).

As a result of this initial meeting, the People's Dialogue began to explore ways in which support could be given to low-income households. From the beginning, there has been a focus on housing, which is a priority need for many of the women living in the informal settlements. A core objective has been to ensure that all activities initiated by the People's Dialogue are those which included women, and within which women could become equal and active participants.[4]

Another important focus of the work of the People's Dialogue was to explore finance options, since it was believed that it is only through such strategies that the autonomy and independence of community organizations could be secured. The conference at Broederstroom had already recognized the need for community-controlled systems of housing finance (Bolnick and Namo 1993). Therefore, it was decided to form housing savings schemes in order to provide a local focus for household saving and organization. As a result of the interest in mobilizing around housing in the South African context with a gender-based division of roles whereby the responsibility for housing improvements and housing finance falls to women, 85 per cent of the membership was made up of women residents.

Between 1991 and 1994, housing savings schemes were established in a growing number of informal settlements in South Africa. Information on savings strategies and housing options was shared between communities either directly through meetings and exchanges or through 'talking newsletters', tapes that told stories about savings initiatives to those who could not read or speak English. By June 1993, there were fifty-eight housing savings schemes and, in 1994, just prior to the first democratic election, these schemes joined together into a national organization, the South African Homeless People's Federation. The Federation serves to formalize the network of autonomous community-based organizations but enables each group to retain its identity and decision-making structure. The organizations are united by a common development approach: namely, that all member organizations are rooted in shack settlements, backyard shacks or hostels; that all organizations are involved in savings and credit managed at the grassroots level by the members themselves; that men are not excluded, but that the vast majority of Federation members are women; and, that all organizations are involved in struggles to attain security of land tenure and affordable housing.

Presently, the sole purpose of the People's Dialogue on Land and Shelter is to support the initiatives of the Federation through fund-raising, administrative support, analysis of information gathered by the Federation on its members and their communities, assistance in negotiations with formal institutions and advocacy for the cause of a people's housing movement in conjunction with Federation representatives.[5] In 1994, a revolving fund was established, supported by R4 million from donor grants and later by R10 million from the South African government, to enable the People's Dialogue and the Federation membership to be directly involved in housing construction. With these funds the People's Dialogue and the Federation have been able to provide loan finance to housing savings schemes to facilitate group housing development. The financial component of this housing programme is known as the uTshani Fund (uTshani = 'grassroots' in Zulu), and it is this Fund that now provides the loan capital for the housing developments of the Federation.

The following sections describe the savings activities of the Federation and the operation of the uTshani Fund, and explore some of their achievements.

Housing savings schemes

The members of housing savings schemes are women who save small amounts of money on a regular basis. These savings are then banked locally and used to provide small-scale finance to members for emergencies and in support of income-generation activities. Federation policy is that people who borrow money from the fund for emergencies are charged 1 per cent per month on the outstanding balance and those who borrow for income generation are charged 2 per cent. All other matters relating to finance management are determined locally and are undertaken by treasurers, book-keepers and collectors drawn from the membership. Table 15.1 indicates that the housing savings schemes are spreading rapidly and that many new members and new schemes are being formed.

A frequent comment from the more sceptical members of a community is that members' savings are minimal in comparison to the total cost of house

Table 15.1 Growth of housing savings schemes

	July *1993*	*July* *1994*	*April* *1995*	*April* *1996*	*April* *1997*
Savings schemes	58	137	198	316	814
Active savers	2,178	7,002	9,627	17,280	43,124
Total savings (R)	34,039	165,023	272,250	452,658	810,000

Source: Data supplied by People's Dialogue.
Note: Figures have not been deflated into constant Rand.

construction. In fact, while the Federation has no illusions that the savings will be adequate to construct houses, it perceives a much more complex relationship between savings and housing finance. The Federation argues that savings mobilize low-income people or, as in its own words, 'We do not collect money, we collect people.' Federation members are mobilized through savings to be self-reliant and to build and run their own organizations with their own resources. Furthermore, savings ensure high levels of participation and mutual interaction. Through investing their scarce resources, members have a material stake in their group and in its planning and decision-making. Savings encourage regular interaction and create strong bonds between members, with the result that the schemes are reliable support systems. Indeed, some of the schemes are the strongest organizations within low-income settlements, not as a coercive force but in terms of empowering their members and creating the space for them to make their own decisions to improve their lives.

Finally, savings create a space for the central participation of women in informal settlements. This is because women are much more interested than men in saving, particularly for housing. As a result the schemes make a material difference to the lives of the most marginalized people in South African society: low-income, homeless women. An important means to achieve this is by savings and loans enabling community organizations to develop the capacity to manage and control finance and to demonstrate this ability to the outside world.

The experience of saving, lending and financial management provides a platform for the acquisition of further development skills. The most important mechanisms to support this learning process are community-to-community exchanges of information, experience and skills, which are organized by the Federation. The case study below describes the work of one of the federation leaders who was responsible for supporting the housing savings schemes during a visit to Gauteng province. She first describes an approach to existing community leaders to ask for a meeting with residents to introduce the Federation, and then a meeting with an established savings scheme.

Case study 1: An Exchange Programme to Vosloorus Ext. 20, 9–11 November 1995

Day one: Meeting with representatives of the Civic and Council Executive. Alinah and I explained the work of the People's Dialogue, the Federation and the Housing Saving Schemes. The representatives understood our outline and promised to call a community meeting.

Day two: The meeting was held for 9:30 a.m. I explained how the People's Dialogue and the Housing Saving Schemes worked and encouraged people to ask questions, which were answered clearly. At this meeting twenty people joined the Federation and started saving with an amount of R100. At 12:30 p.m., Auntie Iris arrived and explained to the older members of Luthando how the affordability forms should be compiled and these were given to them at a later meeting at 6 p.m. They were so excited and the older members started a door-to-door campaign to upgrade older members and they themselves called a meeting for the following day at 9 a.m.

Day Three: We started checking members' savings books and also the bank book and treasurer's books to see if they were in order. We noticed that R714 was outstanding. When I asked about the money, I was told that it had been lent to other members and that it was being collected. The members who borrowed the money said that they would repay it as soon as possible.

From all the exchanges that I'd organized this was the most powerful that I had come across. The reason for this was that the members were very organized and those who had lost interest were upgraded. To date, they have saved R3,200, number 100 members and hold meetings every Wednesday at 5 p.m. They also promised that, by December, they would save more. Alinah and I thought that they should start with their development plans quite soon because their housing savings scheme was not very strong but they have shown commitment to and interest in the People's Dialogue and the South African Homeless People's Federation.

Source: Report by Benedictor Mahlangu, Housing Savings Scheme Convenor, Gauteng.

The training that takes place within community exchanges covers the spectrum of skills needed to consolidate community-managed housing and neighbourhood development. There are four main components to the training programme: a financial component whereby members learn basic financial management, including book-keeping and banking; training in how to map existing houses and local facilities, and gather basic socio-economic data about the community; a modelling process where members collectively design the house of their dreams, consider issues of affordability and amend the designs; and training in construction techniques once loans have been obtained and an acceptable design identified.

From 1991, members of the Federation demonstrated energy, initiative, skill, and experience. However, they lacked the material resources to transform their situation. In 1993, therefore, it was decided that the only way around this problem was for People's Dialogue to assist the Federation in becoming directly involved in managing its own finances and the uTshani Fund was formed.

The uTshani Fund

The uTshani Fund makes finance available directly to housing savings schemes. In order to apply, a scheme has to provide affordability assessments for the potential borrowers and construction plans. Schemes which are new to housing loans normally receive finance for a group of ten members with security provided through 'peer pressure' both within the local network of the housing savings scheme and the broader Federation. Further lending to each scheme is dependent on the performance of existing loans and, because the uTshani Fund is a 'revolving' fund, loan repayments are the source of further loans. The affordability assessments enable the members' monthly repayments to be worked out and uTshani Fund staff determine the total amount which can be borrowed over a fifteen-year period at a simple interest rate of 1 per cent per month. If necessary, the housing savings scheme then adjusts its house designs to build within this amount. One month's grace period is given after the loan is granted. The members then make monthly repayments to their housing savings schemes with the treasurers collecting loan repayments along with savings. Each scheme makes a single monthly repayment to the uTshani Fund.

The final decision on whether or not a loan should be granted rests with the Governing Body, which consists of Federation members from each of the regions in which savings schemes are active. Once a loan has been agreed, the uTshani Fund advances the money to the savings scheme, which then distributes the funds to members in the form of construction materials (not cash). Great emphasis is placed on simplicity and transparency, and schemes are encouraged to have weekly and bi-weekly meetings with open access to information. Book-keeping, release of funds, distribution of construction materials and management of the construction process are undertaken by the members, supported by other Federation groups in the course of regular exchange programmes. The second case study is an extract from a report written by a uTshani Fund convenor following a visit to the Southern Cape.

Case study 2: uTshani Fund Report, visit to Mosselbay and Bossiesgif, 8 August 1995

Mossel Bay members had a special meeting for their problems on Sunday evening at 6 p.m. On Monday morning, I met book-keepers from the Vusisizwe

---

and Imizamo Yethu Housing Savings Scheme. In the afternoon, I met the treasurers and collectors to work out loan interest and deposits.

On Tuesday, I clarified the uTshani Fund system. Members were not familiar with the concept of deposits so I explained very clearly how important these were as security and how, should a member have a problem with a monthly repayment, the uTshani Fund could take from the deposit to pay for that particular month, the same as daily savings.

If a member has enough money in her savings book, she can borrow her deposit from the housing savings scheme. I made it clear that it is not compulsory for a member to save a set amount; they could save more if they wished and they could put even more on deposit if they liked.

It appeared to me that Vusisizwe Housing Savings Scheme in Mossel Bay was a building centre and not a housing saving scheme because they just focus on construction and totally forget about daily savings. I motivated them and they showed great interest in daily savings and deposits although some of them did not even know where their savings books were. They promised to pay deposits in terms and collect it in the BIT [Building, Information and Training] centres account. Then, they would transfer all deposits to the uTshani Fund account.

Source: Report by Florence September, uTshani Fund convenor.

To date, the uTshani Fund has given 795 loans at an average size of R8,920 per household (members can receive a maximum loan of R10,000). Most of the finance has been invested in finished houses of between 50 and 64 square metres (typically with two bedrooms, a living area, kitchen and bathroom) – considerably better than most houses acquired through the capital subsidy programme.

LESSONS FROM THE EXPERIENCE OF PEOPLE'S DIALOGUE

Chapter 2 identified several potential development contributions that may arise as a result of innovative housing finance programmes. In many cases, these contributions are rarely stated as clear outputs in formalized project planning frameworks and are often only articulated in discussions about why programmes were initiated or discussions about programme impacts. This section presents a reflective discussion of the achievements of the People's Dialogue and Federation in South Africa following the structure outlined in Chapter 2.

Material improvements

The need for investment in housing improvement in South Africa is self-evident. Many Federation members are living in self-built shacks of wood and corrugated iron sheets with inadequate protection against the weather and a need for frequent repairs. One survey undertaken by the Federation showed that, in thirty-five settlements between April 1994 and April 1995, the government had neither built housing nor supported the construction process, while the people had constructed 9,500 shacks mostly without (or with only limited) technical advice and finance.

Table 15.2 summarizes the growth in loans given by the uTshani Fund and the number of completed houses. The improved housing has made a clear difference to both the physical comfort and self-respect of those with new houses. The maximum size loan of R10,000 (also the most popular loan) is sufficient to purchase the materials and employ a skilled builder to assist the household construct a 54 square metre two-bedroom house with a living room, kitchen and a corrugated iron roof. The same house built by a contractor would cost an estimated R37,000 (People's Dialogue 1995). Furthermore, the repayments of R120 per month on a R10,000 loan (assuming an interest rate of 12 per cent per annum and a fifteen-year term) are affordable by households with an income of R700 a month (currently the Federation average). For those households that manage to obtain a capital subsidy of R7,500 from the government, the loan finance is reduced to R2,500.[6]

Table 15.2 Finance and house completions from the uTshani Fund

	June 1995	*January 1996*	*June 1996*	*January 1997*
Loans	98	130	245	720
Houses occupied	4	38	97	475

Source: Data supplied by People's Dialogue.

Although such housing can be affordable, the housing development process has had to overcome other obstacles. In one settlement in Plettenburg Bay (Southern Cape), the savings scheme entered into a conflict with the local authority after developers had been granted contracts to build housing with subsidy finance. The savings scheme was already weak, having had difficulty undertaking construction with the loans from the uTshani Fund. In addition, its original leader had accepted the offer of a contractor house and the saving scheme's tools were lost when a hut caught fire. With the support of Federation members, the women improved their construction techniques and they have now completed the ten houses for which they received loan finance. In addition, the women negotiated with the local authority project manager for compensation against the loss of the tools and several members have fully paid off their loans by using the capital subsidies. With the 60 square metre houses now complete, other residents in the settlement who took up the 30 square metre developer houses are anxious to find out if they can obtain loan finance from the uTshani Fund in order to extend their homes.

Income generation and local economic development

The Federation has sought to use the savings schemes to support local economic development through two different strategies, neither of which has been very successful. The first strategy has been to encourage the schemes to lend to members for income generation activities. The second has been to try to establish construction material production related to housing development.

The availability of finance for income generation has come up against the reluctance of women to take out further loans. This reluctance is probably related to the nature of the informal economy in South Africa which Rogerson (1996) shows to exhibit only rarely the characteristics of long-term growth. Moreover, there is a tendency for women to be more prevalent in the areas where there is less potential for growth. Overall, and relative to the Asian economies with which the Federation has undertaken exchange programmes and learnt about combining housing and income generation lending, national economic growth rates in South Africa are low and far below the 3 per cent recorded for India or the 6 per cent in Thailand.

The Federation has made several exploratory attempts to initiate construction material production, especially tile- and brick-making. While important technical skills have been learnt and production successfully undertaken, this has not proved to be a commercial venture. In most cases, it remains cheaper for the Federation to purchase the materials that it needs from manufacturers. The one exception has been a co-operative housing development at Victoria Mxenge, Cape Town where many of the core members (over 80 per cent women) have been enthusiastic about brick-making and the tasks involved in construction and the manufacture of materials are shared between group members.

Local networks and women's participation

One of the acknowledged benefits of the People's Dialogue/Federation approach is the strengthening of local groups (particularly women's groups) and the encouragement of informal networks that can make such a major difference in reducing the vulnerability faced by those with low incomes. In South Africa, the focus on savings has resulted in high levels of participation by women. As People's Dialogue explain:

> Because the focus has been placed on housing, with a particular stress on savings for housing, women who generally feel a greater need for decent secure housing tended to play a leading role. Men, as typical organizational leaders, have been willing to create the space for women's central participation because saving for housing is regarded as 'a women's skill'.
>
> Women's central participation in the Federation is a practical issue. The process through which the poor and excluded can obtain housing

is difficult. Inevitably those who are most committed to improved housing will come to the fore, it is this non-random social selection process which has resulted in the central participation of women.

(People's Dialogue 1995).

The high profile of women represents a significant change from the situation that prevailed within most community organizations and NGOs prior to the establishment of the Federation. At that time, meetings were dominated by men and discussions had little to do with the practical problems faced by most women (Bolnick and Baumann 1995). At the Broederstroom conference that launched the People's Dialogue, for example, over 60 per cent of the delegates were men, but women now hold over 60 per cent of leadership positions in the Federation and form 85 per cent of the membership (People's Dialogue 1995).[7] Savings scheme meetings are dominated by women's voices representing their concerns, and it is now women's understanding that informs the discussions about the strategies that are required to address the problems that women face. At one community meeting in Kleinskool in January 1996, the audience of over one hundred residents was primarily made up of women and all the presentations of the Federation initiatives and strategies were made by women.

The domination by women has not been confrontational for the housing savings schemes and nor, in general, has it harmed their relationship with other local organizations. There are probably good reasons for this: housing is seen as a woman's responsibility and the savings schemes start with very small financial contributions. Nevertheless, as women have come together, developed and consolidated their knowledge and skills, there have been inevitable tensions with the more conventional and predominantly patriarchal community organizations, accustomed to controlling resource flows and dominating development. Over time, however, many of these organizations have come to see the schemes as enormous assets in their community and, instead of seeking to undermine them, have understood that by supporting the schemes they can maximize the development opportunities of their communities as a whole.

The strength of women within savings schemes has meant that they have dominated the exchange programmes between settlements. As a result, the women have become more exposed to other experiences and been encouraged to take on leadership roles. These informal networks have provided a basis for skill sharing and learning, particularly in the areas of financial management, construction and public speaking: women who were previously incapable of articulating their needs in community meetings now challenge cabinet ministers in public.

The savings schemes themselves have been of direct benefit to women through the provision of emergency funds. Procedures for application and approval are informal and flexible, and are mediated by need and not by rules. Some savings schemes interpret 'emergencies' very broadly to include loans related to illness and death (the most frequent trigger); money to pay for medicines for children or transport to hospitals; to cover funeral expenses; or to repay arrears on housing loans to the uTshani Fund. In almost all cases emergency loans are to individual families, but there have been at least two occasions in which savings schemes released money to groups when shacks had been destroyed or damaged by fire. At a collective level, schemes have also developed networks among themselves to cover shortfalls when loan demand exceeds the level of savings.

Interaction with external agencies

The activities around savings and house construction have been effective in drawing external agencies into contact with the community leadership and in enabling a renegotiation of traditional relationships between professionals and residents at both a national and local level.

The speed with which the People's Dialogue and Federation have been able to elicit a national response has been determined by two factors: the decision to federate the savings schemes and the stated aim of the new South African government to assist people-centred development. Creating a federal structure has allowed the settlement-to-settlement learning process to strengthen local consciousness and confidence, and community-to-community exchanges to address the consequences of poverty and apartheid (People's Dialogue *et al.* 1994). The linking of settlements has reduced the anonymity of women and brought an increasing recognition (and self-recognition) of their skills and experience. Women are increasingly aware of what they can do and external agencies are increasingly willing to work with them.

These factors notwithstanding, the speed with which the Federation has been able to play an acknowledged role within the 'stake-holder' style politics of present day South Africa is impressive. Federation representatives now sit on the National Housing Board, on national and provincial Housing Support Task Teams, the National Housing Finance Corporation, and run joint working groups with the Departments of Housing and Land Affairs. In most cases, the Federation representatives are women. Three examples serve to illustrate the nature of the Federation's engagement with national government ministries: negotiations to gain financial support for the uTshani Fund, the access of Federation members to the capital subsidies and support for legal land tenure.

Gaining government support for the uTshani Fund followed a meeting with Joe Slovo, then Minister of Housing, in June 1994 to introduce him to the principles and practices of the housing saving schemes. Several months after this first meeting, in October 1994, the Minister responded with a commitment of R10 million to the uTshani Fund, but despite the evident need for these funds (the Federation had about 8,000 active savers at the time) there was a long delay before they were finally received in January 1996. One reason for the delay appeared to be a reluctance by the South African government to give an open financial commitment to a process that was perceived to lack professional financial controls.[8] Consequently, the uTshani Trust was established, with trustees drawn in equal proportion from government and the Federation, and it was agreed that the release of funds to the Federation would need the approval of the majority of the Trustees. Once the R10 million has been drawn down (and repaid), the funds would be passed over to the Federation and the only obligation to the government would be an annual audit and report on activities.

The second example concerns attempts by the Federation and the People's Dialogue to gain access to the capital subsidy programme. A major problem here is that the Federation does not fit neatly into the not-for-profit categories developed by the Department of Housing and it does not wish to become a housing association (renting properties to those in need) nor a housing co-operative (with collective ownership of land and housing). Lacking a specific category, members had to make individual applications with the attendant bureaucracy this implied. The absurdity of a situation whereby subsidies could not be obtained directly by the beneficiaries was eventually recognized by those with responsibility for the subsidy scheme and by others in the government. However, because distribution takes place at a provincial level, it was not until August 1996, more than a year after Federation members started to demonstrate their repayment record for the housing loans, that progress was made when the Minister for Housing for the Eastern Cape pledged 1,600 subsidies to Federation members.

One of the major problems faced by Federation members in obtaining subsidies has been the condition that recipients prove legal tenure. The Federation and People's Dialogue have been in discussions with the Department of Land Affairs to try to address the issue of urban land tenure. Eleven test sites were identified in which, it was agreed, the Federation would try to obtain security of tenure through existing channels and the Department would seek to learn from their experiences and better understand the obstacles which prevent legal tenure from being secured. In return for this action-research, the Department gave a commitment that the pilot communities obtained legal tenure.

At a national and provincial level, therefore, the Federation leadership have gone some way to securing some movement in policies designed to support low-income households. While progress has been made, changes have been slow. The terms by which the Federation will be able to obtain capital subsidies are still not clear. With respect to the test sites seeking land tenure, two years on, not one of the pilot communities has received title deeds.

The Federation's influence has probably been more significant at the local level where it has challenged the conventional approach of external agencies, be they NGOs or private contractors, in South Africa's informal settlements. Before, the standard procedure was for external agents to work through the local civic organizations to obtain a degree of legitimacy in relations with the community. Civic leaders (mostly men) would act as intermediaries between the community and the external agency, and communication between the leaders and residents would be poor with most people (and especially women) excluded from having any effective influence on, let alone control over, projects. As a result, in the case of subsidies for example, funds would often be misallocated either by allowing the external agency to charge for management services (even though these could be more effectively done by local organizations), by investing in non-priority services (especially for women), by not making the maximum use of the resident labour force, or by insisting on over-specified technical standards.

The training process around the housing savings schemes has provided the members with the knowledge to intervene more effectively more effectively in community development. The socio-economic surveys and mapping exercises mean that local Federation members probably know more about their settlement than any other agency and they have gained the ability to present this information in a form that professionals can understand. The financial training and management of savings has familiarized members with basic book-keeping skills and given them the confidence to deal with financial issues. Finally, and most importantly, the strength of informal networking together with visits to other settlements have given the local leadership the confidence that is needed to deal with, generally, male, and white, professionals.

The consequences of this alternative can be seen in Piesang River, an informal settlement outside Durban which has had an active housing savings scheme for several years. The settlement was part way through an improvement programme which was concentrating on roads, water reticulation and electricity when the people's priority was housing. It was evident to the scheme representatives, therefore, that the programme was wasting scarce resources and they approached the NGO to request changes. When it became evident that the NGO was not prepared to amend its plans, the representatives insisted and, eventually, the NGO was forced to pull out of the project. While the delay to service installation is clearly detrimental for the local population, the settlement has retained the subsidy finance for infrastructure to be used at a later date.

CONCLUSION: LEARNING FROM EXPERIENCE AND LESSONS FOR THE FUTURE

This chapter has considered the recent experience of housing finance by a South African NGO and people's organization that has prioritized the involvement of women. It has sought to show how housing finance has been an effective mobilizing tool to draw women into an organization that is flexible enough to respond to their multiple needs. The housing savings schemes have sought to strengthen the position of the poorest settlements *vis-à-vis* more powerful external agencies, and to support some of the weakest groups within such settlements. Unlike many current finance initiatives, the concern is with social development objectives more than with achieving solely greater efficiency in financial markets and providing commercial borrowing opportunities to low-income people. Through bringing the savings schemes together, a powerful national organization has been created that is able to negotiate with government and other external groups to ensure that policies and programmes better reflect the priorities of low-income urban residents.

It is evident that the activities of People's Dialogue and the Federation have already managed to make a significant improvement to the lives of hundreds and possibly thousands of low-income households living in informal settlements in South Africa. It is clear that the quality of the housing built with the loans is significantly better than that available from private contractors. Furthermore, the Federation has gone to considerable lengths to maintain equity considerations at the forefront of the housing finance programme. Thus, while acknowledging that loans have been offered to the slightly better-off households, it has required (with moderate success) that half of all loans should be small and therefore affordable to those with lower incomes.[9] Most housing loans are given to women, reflecting their overwhelming participation in housing savings schemes. Although there are no central records (these data are managed at the scheme level), which makes it impossible to assess whether loans have gone to female-headed households, the bias towards larger loans suggests that recipient households have dual incomes.[10]

The scaling up of the Federation's operations has been impressive. As of 1996, Federation members had completed only a few houses but, by 1997, the uTshani Fund was managing several hundred loans and the Federation had 40,000 members. More importantly uTshani had already raised the capital for 2,000 R10,000 loans which, assuming that more members than previously receive capital subsidies and thus reduce the average loan size to R2,500, means that the Fund may find itself managing four times as many loans in the next few years. This scaling-up process, however, has faced constraints for two reasons. First, difficulties in the expansion of the technical support for house construction, in this case the training programme to assist members to prepare for housing construction and introduce the Federation to residents within the settlements, has made the rapid increase in loans difficult. Second, South Africa is continuing to undergo rapidly changing policies, political realignments and institutional changes which have produced an environment in which there is little evident stability and many an illusory opening.

The shape of the future relationship with the South African government is important as there is a real danger that the Federation may become over-dependant on external support. If the programme is to reach a size that enables the Federation significantly to address the housing needs of the urban poor, further government support will be required. Yet, if this support is too generous, it may provoke communities to do less for themselves and, without a local organization strengthened by frequent collective tasks and a federal structure to unite groups into a national force, it may be more difficult in future to secure a long-term commitment from government. Moreover, dependence on the capital subsidy, which is set to fall in real terms due to inflation, might provoke bitter and protracted struggles with government to gain access to funds.

In a context in which many groups have argued that either the state or the market should be the main provider of improved housing, the People's Dialogue and Homeless People's Federation have sought to demonstrate that a development process controlled by local residents is both an efficient use of resources and an effective strategy to strengthen organizations that have the potential to address many other development needs. Of course, it is too early to judge the success or otherwise of this strategic decision but it is already evident that the Federation has succeeded in drawing government finance down to support the housing development process of the people, and that without the Federation many changes to government policy would not have taken place.

NOTES

1 The government has stated that the value of the subsidy will not be adjusted for inflation despite the increasing construction costs.
2 In its initial methodology and strategy, the company modelled itself on the Grameen Bank and received advice from Mohammed Yunus, the founder of the Grameen Bank.
3 This figure did fall to 70 per cent as women increasingly applied for housing loans.
4 Broederstroom resulted in a women's forum that continued to meet for some months following the conference.
5 Negotiations with government in 1995 and 1996 attempted to obtain direct access to the capital subsidy for Federation members. In September 1996, the government of Eastern Cape indicated its willingness to offer subsidies directly to Federation members.
6 Inevitably, finance is most easily affordable by the richer households who also want the largest loans. The Governing Body has attempted to redress this by requiring that at least half of the households within each savings group to receive funds falls into the lower rather than the upper size of loans. In practice, however, 80 per cent of loans are maximum size loans.
7 Recent leadership changes brought a further two women onto the Federation's major decision-making body.
8 The advance of funds was not assisted by the death of Joe Slovo and his replacement by Sankie Mthembi-Mahanyele.
9 There have been a few attempts to design cheaper housing such as a 28 square metre house in Piesang River that was constructed for R5,000, although most loans continue to be for R10,000.
10 It is possible that extra incomes are provided by children or siblings which go under-recorded in the surveys.

BIBLIOGRAPHY

Aasland, A. (1996) 'Russians outside Russia: the new Russian diaspora', in G. Smith (ed.) *The Nationalities Question in the Post Soviet States*, London: Longman.

Abbott, J. (1996) *Sharing the City*, London: Earthscan.

Abrams, C. (1966) *Housing in the Modern World: Man's Struggle for Shelter in an Urbanizing World*, Cambridge: MIT Press

Abugre, C. (1993) 'When credit is not due: financial services by NGOs in Africa', *Small Enterprise Development*, 4, 4: 24–33.

Ackerly, B.A. (1995) 'Testing the tools of development: credit programmes, loan involvement, and women's empowerment', *IDS Bulletin*, 26, 3: 56–68.

Adams, D.W. (1992) 'Taking a fresh look at informal finance', in D.W. Adams and D.A. Fitchett (eds) *Informal Finance in Low-income Countries*, Boulder: Westview.

Adams, D.W. and Fitchett, D.A. (eds) (1992) *Informal Finance in Low-income Countries*, Boulder: Westview.

Adams, D.W. and Ghate, P.B. (1992) 'Where to from here in informal finance?' in D.W. Adams and D. A. Fitchett (eds) *Informal Finance in Low-income Countries*, Boulder: Westview.

Adams, D.W. and von Pischke, J.D. (1992) 'Microenterprise credit programs: déjà vu', *World Development*, 20, 10: 1463–70.

Addae-Dapaah, K. and Leong, K.M. (1996a) 'Housing finance for the ageing Singapore population: the potential of the home equity conversion scheme', *Habitat International*, 20, 1: 109–20.

—— (1996b) 'Housing finance for the ageing Singapore population: the potential of creative housing finance schemes', *Habitat International*, 20, 4: 625–34.

Albee, A. and Gamage, N. (1996) *Our Money Our Movement: Building a Poor People's Credit Union*, London: Intermediate Technology Publications.

Alonzo, A. (1994) 'The development of housing finance for the urban poor in the Philippines: the experience of the Home Development Mutual Fund', *Cities*, 11, 6: 398–401.

Andrusz, G. (1992) 'Housing reform and social conflict', in D. Lane (ed.) *Russia in Flux: The Political Consequences of Reform*, Aldershot: Edward Elgar.

Anzorena, J. (1993) 'Supporting shelter improvements for low-income groups', *Environment and Urbanization*, 5, 1: 122–31.

—— (1996) SELAVIP Newsletter (Latin American and Asian Low Income, Housing Service), April, Cebu City: SELAVIP.

Appleton, S. (1996) 'Women-headed households and household welfare', *World Development*, 24, 12: 1811–27.

Ardener, S. and Burman, S. (1995) *Money-go-rounds: The Importance of Rotating Savings and Credit Associations for Women*, Oxford: Berg Publishers.

Arrossi, S., Bombarolo, F., Hardoy, J.E., Mitlin, D., Pérez Coscio, L. and Satterthwaite, D. (1994) *Funding Community Initiatives*, London: Earthscan.

Asthana, S. (1994) 'Integrated slum improvement in Visakhapatnam, India: problems and prospects', *Habitat International*, 18, 1: 57–70.

Baken, R.J. and van der Linden, J. (1993) '"Getting the incentives right": banking on the formal private sector', *Third World Planning Review*, 15, 1:1–22.

Ball, M. (1990) *Under One Roof: The International Financial Revolution and Mortgage Finance*, Harvester Wheatsheaf, Hemel Hempstead.

Bamberger, M. (1982) 'The role of self-help housing in low-cost shelter programmes for the Third World', *Built Environment*, 8, 2: 95–107.

Bamberger, M. and Harth-Deneke, A. (1984) 'Can shelter programmes meet low-income needs? The experience of El Salvador' in G. Payne (ed.) *Low-income Housing in the Developing World: The Role of Sites and Services and Settlement Upgrading*, Chichester: John Wiley & Sons.

Banco de México (1987) *INFONAVIT: Análisis del sistema crediíico de INFONAVIT y propuesta para su modificación*, México DF: Banco de México.

Bank of Botswana (1989) *Housing Finance Institutions and Resource Mobilisation for Housing in Botswana*, Gaborone: Bank of Botswana.

Baross, P. and Mesa, N. (1986) 'From land markets to housing markets: the transformation of illegal settlements in Medellín', *Habitat International*, 10, 3: 153–70.

Bascom, J. (1993) 'The peasant economy of refugee resettlement in Eastern Sudan', *Annals of the Association of American Geographers*, 83, 2: 320–46.

Bennett, L.M. and Cuevas, C.E. (1996) 'Sustainable banking with the poor', *Journal of International Development*, 8, 2: 145–52.

Bennett, L.M., Goldberg, M. and Hunte, P. (1996) 'Ownership and sustainability: lessons on group based financial services from South Asia', *Journal of International Development*, 8, 2: 271–88.

Berger, M. (1989) 'Giving women credit: the strengths and limitations of credit as a tool for alleviating poverty', *World Development*, 17, 7: 1017–32.

Betancur, J.J. (1995) 'Spontaneous settlements in Colombia: from opposition to reluctant acceptance to . . . again opposition', in B.C. Aldrich and R.S. Sandhu (eds) *Housing the Urban Poor: Policy and Practice in Developing Countries*, London: Zed.

Blumberg, R.L. (1995) 'Gender, microenterprise, performance and power: case studies from the Dominican Republic, Ecuador, Guatemala, and Swaziland', in C.E. Bose and E. Acosta-Belén (eds) *Women in the Latin American Development Process*, Philadelphia: Temple University Press.

Bohman, K. (1984) *Women of the Barrio: Class and Gender in a Colombian City*, Stockholm: University of Stockholm.

Boleat, M. (1987) 'Housing finance institutions', in L. Rodwin (ed.) *Shelter, Settlement and Development*, Boston: Allen & Unwin.

Bolnick, J. and Baumann, T. (1995) 'International grassroots networking yields a handsome dividend in the new South Africa', *Trialog*, 47: 7–11.

Bolnick, J. and Namo, I. (1993) 'The People's Dialogue on land and shelter: community driven networking in South Africa's informal settlements', *Environment and Urbanization*, 5, 1: 91–110.

Bond, C. (1979) *Women's Involvement in Agriculture in Botswana*, Gaborone: Government Printer.

Bond, P. (1990) 'Township housing and South Africa's "financial explosion": the theory and the practice of financial capital in Alexandria', *Urban Forum*, 2, 1: 39–67.

Bornstein, D. (1996) *The Price of a Dream: The Story of the Grameen Bank and the Idea That is Helping the Poor Change Their Lives*, New York: Simon & Schuster.

Bouman, F.J.A. (1995) 'Rotating and accumulating savings and credit associations: a development perspective', *World Development*, 23, 3: 371–84.

Bouman, F.J.A. and Hospes, O. (1994) 'Financial landscapes reconstructed', in F.J.A. Bouman and O. Hospes (eds) (1994) *Financial Landscapes Reconstructed: The Fine Art of Mapping Development*, Boulder: Westview Press.

Bratton, M. (1989) 'The politics of government-NGO relations in Africa', *World Development*, 17, 4: 569–87.

—— (1990) 'Non-governmental organizations in Africa: can they influence public policy?' *Development and Change*, 21, 1: 87–118.

Brown, B. (1980) *Women's Role in Development in Botswana*, Gaborone: Government Printer.

—— (1983) 'The impact of male labour migration on women in Botswana', *African Affairs*, 82: 367–88.

Buckley, R.M. (1996) *Housing Finance in Developing Countries*, Basingstoke: Macmillan.

Buckley, R.M. and Mayo, S. (1989) 'Housing policy in developing economies: evaluating the macroeconomic impacts', *Review of Urban and Regional Development Studies*, 2: 27–47.

Burgess, R. (1982) 'Self-help housing advocacy: a curious form of radicalism. A critique of the work of J.F.C. Turner', in P.M. Ward (ed.) *Self-help Housing: A Critique*, London: Mansell.

—— (1985) 'The limits of state self-help housing programmes', *Development and Change*, 16, 2: 271–312.

—— (1986) 'The political integration of urban demands in Colombia', *Boletín de estudios latinoamericanos y del Caribe*, 41: 29–52.

Buvinic, E., Lycette, M. and McGreevey, W. (1983) *Women and Poverty in the Third World*, Baltimore: Johns Hopkins University Press.

Cabannes, Y. (1997) 'From community development to housing finance: from Mutiroes to Casa Melhor in Fortaleza, Brazil', *Environment and Urbanization*, 9, 1: 31–58

Cámara Nacional de la Industria de la Construcción (CNIC) (1996) *Estudio sobre la crisis en el sector vivienda*, México DF: CNIC.

Carroll, T. (1992) *Intermediary NGOs: The Supporting Link in Grassroots Development*, Connecticut: Kumarin Press.

Central Statistics Office (1991) *1991 Population and Housing Census: Administrative/Technical Report and National Statistical Tables*, Gaborone: Government Printer.

Cernea, M.M. (1995) 'Understanding and preventing impoverishment from displacement: reflections on the state of knowledge', *Journal of Refugee Studies*, 8, 3: 245–64.

Chambers, R. (1995) 'Poverty and livelihoods: whose reality counts?' *Environment and Urbanisation*, 7, 1: 173–204.

Chant, S. (1991) 'Women's work and household change in the 1980s,' in N. Harvey (ed.) *Mexico: Dilemmas of Transition*, London: Tauris.

—— (1997) *Women-headed Households: Diversity and Dynamics in the Developing World*, Basingstoke: Macmillan.

Comité Preparatorio de México Conferencia Habitat II (1996) *Informe Nacional*, mimeograph, México DF.

Community Mortgage Bulletin (1995a) *Report Published by the National Congress of the Community Mortgage Programme Originators and Social Development Agencies for Low-income Housing*, May–June, The Philippines.

Community Mortgage Bulletin (1995b) *Report Published by the National Congress of the Community Mortgage Programme Originators and Social Development Agencies for Low-income Housing*, July–October, The Philippines.

Connolly, P. (1993) 'The go-between: CENVI, a Habitat NGO in Mexico City', Environment and Urbanization, 5, 1: 68–90,

Copestake, J. (1996) 'NGO-state collaboration and the new policy agenda: the case of subsidized credit', *Public Administration and Development*, 16, 1: 21–30.

Coulomb, R. (1990) 'México, la política habitacional en la crisis, viejas contradicciones, nuevas estrategias y actores emergentes,' in Habitat International Coalition (ed.) *Políticas habitacionales, y ajustes de la economía en los 1980s*, México DF, Habitat International Coalition.

Central Provident Fund (CPF) (1995) *Annual Report*, Singapore: CPF.

Corporation for Research Development and Education (CORDE) (1994) *CORDE Shelter Study*, Gaborone: CORDE.

Cruz, L.F. (1994) 'Fundación Carvajal: the Carvajal Foundation', *Environment and Urbanization*, 6, 2: 175–82

Dandekar, H.C. (ed.) (1992) *Shelter, Women and Development: First and Third World Perspectives*, Ann Arbor: George Wahr.

Dandekar, V.M., Sukhatme, P.V., Rangarajan, C., Vaidyanathan, A., Radhakrishna, R., Guhan, S., Tendulkar, S.D., Ray, S.N. and Hashim, S.R. (1993) *Report of the Expert Group on Estimation of Proportion and Number of Poor*, Perspective Planning Division. New Delhi: Government of India.

Daniell, J. and Struyk, R. (1997) 'The evolving housing market in Moscow: indicators of housing reform', *Urban Studies*, 34, 2: 235–54.

Datta, K. (1994) 'The rental market for low income housing: a case study of Gaborone, Botswana', unpublished doctoral dissertation, University of Cambridge.

—— (1995) 'Strategies for urban survival? Women landlords in Gaborone, Botswana', *Habitat International*, 19, 1: 1–12.

—— (1996) 'Women owners, tenants and sharers in Botswana', in A. Schlyter (ed.) *A Place to Live: Gender Research on Housing in Africa*, Uppsala: Nordic Africa Institute.

—— (1998) 'Gender, labour markets and migration in and from Botswana', in D. Simon (ed.) *Reconfiguring the Region: South Africa in Southern Africa*, Oxford: James Currey.

Denaldi, R. (1994) 'Viable self-management: the FUNACOM housing programme of São Paulo municipality', *IHS Working Papers* no. 9, Rotterdam: Institute of Housing Studies.

Department of Statistics (1996) Singapore 1965–95: Statistical Highlights – A Review of 30 Years' Development, Singapore: Ministry of Trade and Industry.

Derkyi, E. (1994) 'Providing affordable and sustainable housing loans to lower income earners: Ghana's experience', paper presented at 2nd Symposium Housing for the Urban Poor, Birmingham.

250

Derrick, J. (1992) 'Cameroon: from oil boom to recession', *Africa Recovery*, 16–21.

Desai, V. (1995) *Filling the Gap: An Assessment of the Effectiveness of Urban NGOs*, Final Report to ODA, Sussex: Institute of Development Studies.

Devas, N. (1983) 'Financing urban land development for low-income housing: an analysis with particular reference to Jakarta, Indonesia', *Third World Planning Review*, 5, 3: 209–25.

Diamond, D.B. and Lea, M. (1992) 'The decline of special circuits in developed country housing finance', *Housing Policy Debate*, 3, 3: 747–77.

—— (1995) 'Sustainable housing for housing: a contribution to Habitat II', *Fannie Mae Office of Housing Research Working Paper*, Washington DC: Fannie Mae.

Dichter, T.W. (1996) 'Questioning the future of NGOs in microfinance', *Journal of International Development*, 8, 2: 259–69.

Dizon, A. (1997) *Supporting Community Level Initiatives to Address Environmental Problems in Third World Cities: Metro-Manila Case Studies*, Quezon City: Urban Poor Associates.

Dwyer, D. and Bruce, J. (1988) *A Home Divided: Women and Income in the Third World*, Stanford: Stanford University Press.

Ebdon, R. (1995) 'NGO expansion and the fight to reach the poor: gender implications of NGO scaling-up in Bangladesh', *IDS Bulletin*, 26, 3: 49–55.

Economic and Social Commission for Asia and the Pacific (ESCAP) (1991) 'Guidelines on community based housing finance and innovative credit systems for low-income households', Bangkok: United Nations, ST/ESCAP/1003.

Edwards, M. (1992) 'Policy priorities and strategies for increased cooperation at the international level between non-governmental organizations, governments, international aid agencies, United Nations agencies and inter-governmental organizations', paper presented at Meeting on Governmental/Non-governmental Cooperation in the Field of Human Settlements, The Hague.

—— (1993) '"Does the doormat influence the boot?": critical thoughts on UK NGOs and international advocacy', *Development in Practice*, 3, 3: 163–75.

Edwards, M. and Hulme, D. (1992a) 'Scaling up NGO impact on development: learning from experience', *Development in Practice*, 2, 2: 77–91.

—— (1992b) 'Making a difference: concluding comments', in Edwards, M. and D. Hulme (eds), *Making a Difference: NGOs and Development in a Changing World*, London: Earthscan.

—— (eds) (1996) *Beyond the Magic Bullet: NGO Performance and Accountability in the Post-Cold War World*, West Hartford: Kumarin Press.

Edwards, S. (1995) 'Why are savings rates so different across countries? An international comparative analysis', *NBER Working Paper* 5097.

Enge, M. (1982) *Women in Botswana: Dependent Yet Independent*, Gaborone: SIDA Publications.

Escalona, A. and Black, R. (1995) 'Refugees in Western Europe: bibliographic review and state of the art', *Journal of Refugee Studies*, 8, 4: 364–89.

Espinosa, L. and López Rivera, O.A. (1994) 'UNICEF's urban basic services programme in illegal settlements in Guatemala City', *Environment and Urbanization*, 6, 2: 9-29.

Eyre, L.A. (1997) 'Self-help housing in Jamaica,' in R.B. Potter and D. Conway (eds) *Self-help Housing, the Poor and the State: Pan-Caribbean Perspectives*, Knoxville and Kingston: University of Tennessee and University of West Indies Press.

Flesis, A. (1985) 'Home ownership alternatives for the elderly', London: Department of the Environment, HMSO.

Galindo Guameros, G. (1962) *Las instituciones de ahorro y préstamo para la vivienda popular*, México DF: Editorial Jus.

García, M. and Frigerio, M. (1994) 'The survival of the mutual aid cooperative in the process of liberalisation and non-intervention of the state', paper presented at 2nd Symposium Housing for the Urban Poor, Birmingham.

Garza, G. and Schteingart, M. (1978) *La acción habitacional del estado mexicano*, México, El Colegio de México.

Government of Andhra Pradesh (GAP) (1994) 'Report of the comptroller and auditor general for the year ended 31 March 1993, no. 1 (commercial)', Hyderabad: Government of Andhra Pradesh.

George, V. (1996) 'How has the NHT done?' in D. Cecile (ed.) *An Assessment of the Impact of NHT Schemes, 1976–1996*, Jamaica: NHT Research Department.

Gibson, A. (1993) 'NGOs and income-generation projects: lessons from the Joint Funding Scheme', *Development in Practice*, 3, 3: 184–95.

Gilbert, A. (1997) 'On subsidies and home ownership: thoughts on Colombian housing policy during the 1990s', *Third World Planning Review*, 19, 1: 51–70.

—— (forthcoming) 'A home forever? residential mobility and home ownership in self-help settlements', *Environment and Planning A*.

Gilbert, A. and Ward, P.M. (1982) 'Low-income housing and the state', in A.G. Gilbert, J.E. Hardoy, and R. Ramirez (eds) *Urbanization in Contemporary Latin America*, Chichester: John Wiley & Sons.

Gilfillian, K. (1978) 'The role of government and NHT in housing', in *Housing for Jamaica: Workshops on Science and Technology*, Jamaica: Kingston Scientific Research Council.

The Gleaner (1975) Text of Prime Minister's statement, October 9th, Kingston.

The Gleaner (1991) 'What is middle-income housing: for whom? 75% earn less than $25,000 per annum', March 16th, Kingston.

Glickman, M. (1988) 'Thoughts on certain relationships between gender, kinship and development among the Tswana of Botswana', *African Studies*, 47, 2: 81–8.

Gobierno del Estado de Jalisco (1995) *Proyecto de programa estatal de suelo y vivienda, 1995–2001*, Guadalajara: Gobierno del Estado de Jalisco

Goetz, A.M. and Gupta, R.S. (1996) 'Who takes the credit? Gender, power, and control over loan use in rural credit programs in Bangladesh', *World Development*, 24, 1: 45–63.

González de la Rocha, M.C. (1994) *The Resources of Poverty: Women and Survival in a Mexican city*, Cambridge MA and Oxford: Basil Blackwell.

Gough, K.V. (1992) 'From bamboo to bricks: self-help housing and the building materials industry in urban Colombia', unpublished doctoral dissertation, University of London.

—— (1996a) 'Home-based enterprises in low-income settlements: evidence from Pereira, Colombia', *Danish Journal of Geography*, 96: 95–102.

—— (1996b) 'Linking production, distribution and consumption: self-help builders and the building materials industry in urban Colombia', *Third World Planning Review*, 18, 4: 397–414.

—— (1996c) 'Self-help housing in urban Colombia: alternatives for the production and distribution of building materials', *Habitat International*, 20, 4: 635–51.

—— (1998) 'House for sale? The self-help housing market in Pereira, Colombia', *Housing Studies*, 13, 2 :149–60.

Greene, M. (1993) *La gestión gubernamental en el sector de vivienda y urbanismo, 1991-1992*, Santiago: Corporación de Promoción Universitaria.

Grown, C.A. and Sebstad, J. (1989) 'Introduction: toward a wider perspective on women's employment,' *World Development*, 17, 7: 937–52.

Habitat International Coalition (HIC/ACHR) (1994) *Finance and Resource Mobilization for Low-income Housing and Neighbourhood Development: A Workshop Report*, Pagtambayayong City: Cebu Foundation.

Habitat International Coalition (Women and Shelter Network) (1991) *Environment and Urbanization*, 3, 2: 82–6.

Habitat International Coalition (1997) *Building the City with the People: New Trends in Community Initiatives in Cooperation with Local Governments*, México DF: HIC.

Hamnett, C. (1994) 'Restructuring housing finance and the housing market', in S. Corbridge, N. Thrift and R. Martin (eds) *Money, Power and Space*, Oxford: Basil Blackwell.

Harvey, C. (1992) 'Botswana: is the economic miracle over?', *IDS Discussion Paper* no. 298, University of Sussex: Institute of Development Studies

Havers, M. (1996) 'Financial sustainability in savings and credit programmes', Development in Practice, 6, 2: 144–50.

Henry, A., Tchente, G. and Guillerme-Dieumegard, P. (1991) *Tontines et banques au Cameroon: les principes de la Société des Amis*, Paris: Editions Karthala.

Heumann, L. and Boldy, D. (1982) *Housing for the Elderly: Planning and Policy Formulation in Western Europe and North America*, New York: St Martin's Press.

Hope, P. (1976) 'Uncensored', *The Gleaner*, 22 May, Kingston.

Housing Development Board (HDB) (1993) 'Social aspects of public housing in Singapore: kinship ties and neighbourly relations: sample household survey 1993', Singapore: HDB, Research and Planning Department.

Hulme, D. and Mosley, P. (1996) *Finance against Poverty: Effective Institutions for Lending to Small Farmers and Micro-entreprises in Developing Countries*, London: Routledge.

Igel, B. and Srinivas, H. (1996) 'The co-optation of low-income borrowers by informal credit suppliers: a credit delivery model for squatter housing', *Third World Planning Review*, 18, 3: 287–305.

Institute of Housing Studies (1992) 'Strategies and practical modalities for increased cooperation in human settlements between government institutions, NGOs and CBOs at the local, State/provincial and national levels', paper presented at Meeting on Governmental/Non-governmental Cooperation in the Field of Human Settlements, The Hague.

INFONAVIT (1987) *Análisis del sistema crediticio de INFONAVIT y propuesta para su modificación*, México DF: INFONAVIT.

—— (1992) *20 Años*, México DF: INFONAVIT.

—— (1996) *Plan de financiamiento de vivienda*, México DF: INFONAVIT.

International Research and Training Institute for the Advancement of Women (INSTRAW) (1995) *Credit for Women: Why Is It So Important?* Santo Domingo: INSTRAW.

Jackson, C. (1996) 'Rescuing gender from the poverty trap', *World Development*, 24: 3, 489–504.

Jain, P.S. (1996) 'Managing credit for the rural poor: lessons from the Grameen Bank', *World Development*, 24, 1: 79–89.

Johnson, S. and Rogaly, B. (1997) *Microfinance and Poverty Reduction*, Oxfam Development Guidelines, Oxford: Oxfam.

Joint Center for Housing Studies of Harvard University (1997) *The State of Mexico's Housing*, Cambridge MA.: Harvard University.

Jones, E., Webber, M. and Turner, M.A. (1987) 'Jamaica shelter sector strategy, phase 1, Final Report to USSAID', unpublished.

Jones, G.A. (1996) 'The difference between truth and adequacy: (re)joining Baken, van der Linden and Malpezzi', *Third World Planning Review*, 18, 2: 243–56.

Jones, G.A., Jiménez, E. and Ward, P.M. (1993) 'The land market boom in Mexico under Salinas: a real-estate boom revisited?' *Environment and Planning A*, 25: 627–51.

Jones G.A. and Ward, P.M. (1994) 'The World Bank's "New" Urban Management Programme: paradigm shift or policy continuity?', *Habitat International*, 18, 3: 33–51.

Kabeer, N. (1991) 'Gender, production and well-being: rethinking the household economy', *IDS Discussion Paper* no. 288, University of Sussex: Institute of Development Studies.

—— (1994) *Reversed Realities: Gender Hierarchies in Development Thought*, London: Verso.

Katsura, H.M., Struyk, R.J. and Newman, S.J. (1989) *Housing for the Elderly in 2010: Projections and Policy Options*, Washington, DC: Urban Institute Press.

Kellett, P. and Garnham, A. (1995) 'The role of culture and gender in mediating the impact of official interventions in informal settlements: a study from Colombia', *Habitat International*, 19, 1: 53–60.

Kim, K.H. (1997) 'Housing finance and urban infrastructure finance', *Urban Studies*, 34, 10: 1597–620.

Kinsman, M. (1983) 'Beasts of burden: the subordination of Tswana women, ca. 1800–1840', *Journal of Southern African Studies*, 10: 39–54.

Klak, T. (1992a) 'Excluding the poor from low-income housing programs: the roles of state agencies and ISAID in Jamaica', *Antipode*, 24, 2: 87–112.

—— (1992b) 'What causes arrears in government housing programs? Perceptions and the empirical evidence in Latin America', *Journal of the American Planning Association*, 58, 3: 336–45.

—— (1993) 'Contextualizing state housing programs in Latin America: evidence from leading housing agencies in Brazil, Ecuador, and Jamaica', *Environment and Planning A*, 25: 653–76.

Klak, T. and Hey, J. (1992) 'Gender and state bias in Jamaican housing programmes', *World Development*, 20, 2: 213–27.

Kosareva, N.B. (1993) 'Housing reform in Russia: first steps and future potential', *Cities*, 10, 3: 198–207.

Koussoudji, S. and Mueller, E. (1983) 'The economic status of female-headed households in rural Botswana', *Economic Development and Cultural Change*, 31, 4: 832–59.

Kua, E.H. (1994) *Ageing and Old Age among Chinese in a Singaporean Urban Neighbourhood*, Singapore: Singapore University Press.

La Direction Nationale du DRGP (n.d.) 'Deuxième recensement général de la population et de l'habitat de la République du Cameroun 1987', Yaoundé: Government of Cameroon.

Larsson, A. (1988) *From Outdoor to Indoor Living: The Transition from Traditional to Modern Low Cost Housing in Gaborone*, Sweden: Walkin & Dalholm.

Lau, K.E. (1992) 'Singapore census of population 1990: households and housing', Singapore: Department of Statistics.

Lea, M.J. (1996) 'Restarting housing finance in Mexico', paper presented at Mexico/US Global Forum, Aspen, Colorado.

Leather, P. and Rose, W. (eds) (1988) *Making Use of Equity in Old Age: Report of a Research Project on the Housing Needs of Elderly Owner-occupiers*, London: Building Societies Association.

Lee, M. (1995) 'The community mortgage program: an almost-successful alternative for some urban poor', *Habitat International*, 19, 4: 529–46.

Lowder, S. (1986) *Inside Third World Cities*, London: Croom Helm.

Mackintosh, S., Means, R. and Leather, P. (1990) 'Housing in later life', Bristol: School for Advanced Urban Studies, University of Bristol.

Macoloo, G.C. (1989) 'The commodification of urban self-help housing in Kenya: an analysis of the nature of the changing production and consumption of building materials in Mombassa', unpublished doctoral dissertation, University of Cambridge.

—— (1994) 'The changing nature of financing low-income urban housing development in Kenya', *Housing Studies*, 9, 2: 281–99.

Malik, T.H. (1994) 'Recent development in housing finance policy in Pakistan', paper presented at 2nd Symposium Housing for the Urban Poor, Birmingham.

Malpezzi, S. (1990) 'Urban housing and financial markets: some international comparisons', *Urban Studies*, 27, 6: 971–1022.

—— (1994) '"Getting the incentives right": a reply to Robert-Jan Baken and Jan van der Linden', *Third World Planning Review*, 16, 4: 451–66.

Mayo, S. (1993) *South African Housing Sector Performance in International Perspective*, paper prepared for the 21st IAHS World Housing Congress, South Africa.

Mazingira Institute (n.d.) *International Seminar on African Urban Management. Country Case Study: Kenya*, Nairobi: mimeograph.

Mehta, M. (1994) *Downmarketing Housing Finance through Community Based Housing Financial System*, New Delhi: Abt Associates.

Mencher, J.P. (1988) 'Women's work and poverty: women's contribution to household maintenance in South India', in D. Dwyer and J. Bruce (eds) *A Home Divided: Women and Income in the Third World*, Stanford: Stanford University Press.

Merrett, S. and Russell, K. (1994) 'Non-conventional finance for self-help housing', *Habitat International*, 18, 2: 57–69.

Microcredit Summit (1997a) *Declaration and Plan of Action*, Washington, DC: Results Education Fund.

—— (1997b) *The Microcredit Summit Report*, Washington, DC: Results Education Fund.

Mills, B. and Sahn, D.E. (1996) 'Life after public sector job loss in Guinea', in D.E. Sahn (ed.) *Economic Reform and the Poor in Africa*, Oxford: Clarendon Press.

Ministry of Finance and Development Planning (1991) *National Development Plan VII*, Gaborone: Government Printer.

Ministry of Local Government and Lands (1978) *Gaborone Growth Study: Final Report*, Gaborone: Government Printer.

Ministry of Local Government, Lands and Housing (1992) *Review of the Self-help Housing Agency*, Gaborone: Government Printer.

—— (1997) 'Review of the National Housing Policy', interim report.

Miraftab, F. (1992) 'Shelter as sustenance: exclusionary mechanisms limiting women's access to housing', in H.C. Dandekar (ed.) *Shelter, Women and Development: First and Third World Perspectives*, Ann Arbor: George Wahr.

—— (1994) *Housing Preferences of Female-headed Households of Low-income Families in Guadalajara, Mexico*, paper presented at the International Seminar on Gender, Urbanisation and the Environment, Nairobi.

Miranda, G. and Frigerio, M. (1994) 'The survival of the mutual aid cooperative in the process of liberalisation and non-intervention of the state' paper presented at 2nd symposium Housing for the Urban Poor, Birmingham.

Mitlin, D. (1996) *Housing Finance and Resource Mobilization: Recent Innovations*, London: IIED.

—— (1997) 'Building with credit: housing finance for low-income households', *Third World Planning Review*, 19, 1: 21–50.

Mitlin, D. and Satterthwaite, D. (1992) *Past Experiences with Cooperation Between Governments and NGOs in the Field of Human Settlements*, paper presented at Meeting on Governmental/Non-governmental Cooperation in the Field of Human Settlements, The Hague.

Mogwe, A. (1992) *Country Gender Analysis: Botswana*, report submitted to SIDA. Mimeo.

Montgomery, R. (1996) 'Disciplining or protecting the poor? Avoiding the social costs of peer pressure in micro-credit schemes', *Journal of International Development*, 8, 2: 289–305.

Morton, H.W. (1987) 'Housing quality and housing classes in the Soviet Union', in H. Herlemann (ed.) *Quality of Life in the Soviet Union*, Boulder: Westview.

Moser, C.O.N. (1992a) 'Adjustment from below: low-income women, time and the triple role in Guayaquil, Ecuador', in H. Afshar and C. Dennis (eds) *Women and Adjustment Policies in the Third World*, Basingstoke: Macmillan.

—— (1992b) 'Women and self-help housing projects: a conceptual framework for analysis and policy-making', in K. Mathey (ed.) *Beyond Self-help Housing*, London: Mansell.

Moser, C.O.N. and Chant, S. (1985) 'The role of women in the execution of low income housing projects draft training manual', *DPU Gender and Planning Working Paper* no. 6, London: Development Planning Unit.

Moser, C.O.N. and L. Peeke (eds) *Women, Human SSettlements and Housing*, London: Tavistock.

Mthembi-Mahanyele, S. (1996) Briefing given by Housing Minister Sankie Mthembi-Mahanyele to ANC media summit, Cape Town, 7 November.

Mumtaz, B.K. (1995) 'Housing finance to meet the needs of Sri Lanka', *Economic Review*, June/July, 12–19.

Munjee, N. (1994) 'Housing finance in development: is there an emerging paradigm for developing countries in Asia?', *Housing Finance International*, 8, 4: 6–10.

Mutua, K., Nataradol, P. and Otero, M. (1996) 'The view from the field: perspectives from managers of microfinance institutions', *Journal of International Development*, 8, 2: 179–93.

National Housing Trust (NHT) (1995) *Annual Report 1994–5*, Kingston: NHT.

National Institute of Urban Affairs (NIUA) (1992) 'The informal finance for urban housing: status and prospects', *Research Study Series* no 47, New Delhi: NIUA.

Newcomer, R.J., Lawton, M.P. and Byerts, T.O. (eds) (1986) *Housing An Ageing Society: Issues, Alternatives and Policy*, New York: Van Nostrand Reinhold.

Nimpuno-Parente, P. (1987) 'The struggle for shelter: women in a site and services project in Nairobi, Kenya', in C.O.N. Moser and L. Peake (eds) *Women, Human Settlements and Housing*, London: Tavistock.

O'Reilly, C. (1996) 'Urban women's informal savings and credit systems in Zambia', *Development in Practice*, 6, 2: 165–9.

O'Reilly, C. and Gordon, A. (1995) 'Survival strategies of poor women in urban Africa: the case of Zambia', *NRI Socio-economic Series* 10, Chatham: Natural Resources Institute.

Okpala, D. (1994) 'Financing housing in developing countries: a review of the pitfalls and potentials in the development of formal housing finance systems', *Urban Studies*, 31, 9: 1571–86.

Ortíz Cañas, E.A. (1994) 'Mujeres del sector informal urbano en El Salvador', *Serie estudios de la mujer* 3. San Salvador: Instituto de Investigación, Capacitación, y Desarrollo de la Mujer.

Ortíz, E. (1995) *FONHAPO, gestión y desarrollo de un fondo público en apoyo de la producción social de vivienda*, México DF: Habitat International Coalition.

Oruwari, Y. (1992) 'The invisible contribution of married women to housing finance and its legal implications: a case study from Port Harcourt, Nigeria', in H.C. Dandekar (ed.) *Shelter, Women and Development: First and Third World Perspectives*, Ann Arbor: George Wahr.

Osondu, I.N. and Middleton, A. (1994) *Informal Housing Finance in Nigeria*, paper presented at 2nd Symposium Housing for the Urban Poor, Birmingham.

Papanek, H. and Schwede, L. (1988) 'Women are good with money: earning and managing in an Indonesian city', *Economic and Political Weekly*, October: 73–84.

Patel, S. and Burra, S. (1994) 'Access to housing finance for the urban poor: institutional innovations in India', *Cities*, 11, 6: 393–7.

Peake, L. (1987) 'Government housing policy and its implications for women in Guyana', in C.O.N. Moser and L. Peake (eds) *Women, Human Settlements and Housing*, London: Tavistock.

People's Dialogue (1993) *Housing Finance Packages for Low-Income Housing in South Africa*, Cape Town: People's Dialogue.

—— (1995) *You See a Monument: We See the Way to Build a Million Houses*, Cape Town: People's Dialogue.

—— (1996) *The South African Homeless People's Federation, People's Dialogue, uTshani Fund and the uTshani Trust*, Cape Town: People's Dialogue.

—— (1997) 'uTshani BuyaKhuluma: a profile of the Western Cape Federation', *People's Dialogue Newsletter*, 7 May.

People's Dialogue and SPARC/NSDF/Mahila Milan (1994) *Regaining Knowledge: An Appeal to Abandon Illusion*, Cape Town and Bombay: People's Dialogue.

Peterson, G. and Klak, T. (1990) *Shelter Sector Mortgage Credit and Subsidy Policy: The National Housing Trust and Caribbean Housing Finance Corporation*, Executive Report to USAID and Jamaican Ministries of Construction and Development, Planning and Production, unpublished.

Pezzoli, K. (1995) 'Mexico's urban housing environments: economic and ecological challenges of the 1990s', in B.C. Aldrich and R.S. Sandu (eds) *Housing and the Urban Poor: Policy and Practice in Developing Countries*, London: Zed.

Planact (1992) *Annual Report 1991–2*, Johannesburg: Planact.

Planning Commission of India (1992) *Report of the Working Group on Finance for the Eighth Plan, 1992–1997*, New Delhi: Planning Commission.

Planning Institute of Jamaica (various) *Economic and Social Survey, Jamaica*, Kingston: PIJ.

Pohlmann, L. (1995) 'Ambivalence about leadership in women's organisations: a look at Bangladesh', *IDS Bulletin*, 26, 3: 117–24.

Porio, E. (1997) (ed.) *Urban Governance and Poverty Alleviation in South-east Asia*, Global Urban Research Initiative in South-east Asia, Ateneo de Manila University.

Prisliha, (1995) 'Government social housing, losing focus?' *SELAVIP Newsletter* (Latin American and Asian Low Income, Housing Service), April. Cebu City: SELAVIP.

Pryer, J. (1993) 'The impact of adult ill-health on household income and nutrition in Khulna, Bangladesh', *Environment and Urbanisation*, 5, 2: 35–49.

Pugh, C. (1994) 'Development of housing finance and the global strategy for shelter', *Cities*, 11, 6: 384–92.

Rahman, M.M. (1994) 'A review of the institutional housing finance situation of the urban areas of Bangladesh', *Third World Planning Review*, 16, 1: 71–85.

Rakodi, C. (1995) 'Housing finance for lower income urban households in Zimbabwe', *Housing Studies*, 10, 2: 199–227.

Rakowski, C. (ed.) (1994) *Contrapunto: The Informal Sector Debate in Latin America*, Albany: State University of New York Press.

Remenyi, J. (1991) *Where Credit Is Due: Income-generating Programmes for the Poor in Developing Countries*, London: Intermediate Technology Publications.

Renaud, B. (1987a) 'Another look at housing finance in developing countries', *Cities*, 4, 1: 28–34.

—— (1987b) 'Financing shelter', in L. Rodwin (ed.) *Shelter, Settlement and Development*, Boston: Allen & Unwin.

—— (1995) 'The real estate economy and the design of Russian housing reforms, part 1', *Urban Studies*, 32, 8: 1247–64.

Republic of Zambia (1993) *Social Dimensions of Adjustment: Priority Survey I, 1991*, Lusaka: Central Statistical Office.

Rhyne, E. and Otero, M. (1992) 'Financial services for microenterprises: principles and institutions', *World Development*, 20, 11: 1561–71.

—— (eds) (1994) *The New World of Micro Enterprise Finance: Building Healthy Financial Institutions for the Poor*, West Hartford: Kumarin Press.

Robertson, A.F. (1991) *Beyond the Family: The Social Organization of Reproduction*, Cambridge: Polity Press.

Robertson, S. (1992) 'Women in the urban economy', in M. Hay and S. Stichter (eds) *Africa South of the Sahara*, London: Longman.

Robinson, M.S. (1996) 'Addressing some key questions on finance and poverty', *Journal of International Development*, 8, 2: 153–61.

Robinson, V. (1995) 'The changing nature and European perceptions of Europe's refugee problem', *Geoforum*, 26, 4: 411–27.

Rodell, M.J. (1983) 'Sites and services and low-income housing', in R.J. Skinner and M.J. Rodell (eds) *People, Poverty and Shelter: Problems of Self-help Housing in the Third World*, London: Methuen.

Rogaly, B. (1996) 'Micro-finance evangelism, "destitute women", and the hard selling of a new anti-poverty formula', *Development in Practice*, 6, 2: 100–12.

Rogers, R. (1992) 'The future of refugee flows and policies', *International Migration Review*, 26, 4: 1112–43.

Rogerson, C.M. (1996) 'Urban poverty and the informal economy in South Africa's economic heartland', *Environment and Urbanization*, 8, 1: 167–181.

Rogge, L. (ed.) (1987) *Refugees: A Third World Dilemma*, New Jersey: Rowman & Littlefield.

Rojas, E. and Greene, M. (1995) 'Reaching the poor: lessons from the Chilean housing experience', *Environment and Urbanization*, 7, 2: 31–49.

Rose, E.A. (1982) *Housing Needs and the Elderly*, Aldershot: Gower.

Rueda, N., Angulo, L., Vila, P. and Barney, G. (1979) *La autoconstrucción de vivienda urbana: fundamentos para un enfoque analítico*, Bogotá: CPU, Universidad de los Andes.

Santana, P. and Casabueñas, C. (1981) 'Hacia una política de vivienda popular en Colombia', in P. Santana (ed.) *La vivienda popular hoy en Colombia*, Bogotá: CINEP.

Satterthwaite, D. (1995) 'The scale and nature of international donor assistance to housing, basic services and other human-settlements related projects', paper presented to the UNU/WIDER Conference on Human Settlements in the Changing Global Political and Economic Processes, Helsinki.

Schmidt R.H. and Zeitinger, C.P. (1996a) 'Prospects, problems and potential of credit granting NGOs', *Journal of International Development*, 8, 2: 241–58.

—— (1996b) 'The efficiency of credit-granting NGOs in Latin America', *Savings and Development*, 20, 3: 353–84.

Schrieder, G.R. and Cuevas, C.E. (1992) 'Informal financial groups in Cameroon' in D.W. Adams and D.A. Fitchett (eds) *Informal Finance in Low-income Countries*, Boulder: Westview.

Secretaría de Desarrollo Urbano y de Ecología (SEDUE) (1992) *Estadística de vivienda*, México DF: SEDUE.

Secretaría de Desarrollo Social (SEDESOL) (1996) *Programa nacional de vivienda, 1995–2000*, México DF: SEDESOL.

Sen, G. and Grown, C. (1987) 'Development alternatives with women for a new era', in G. Sen and C. Grown (eds) *Development, Crisis and Alternative Visions: Third World Women's Perspectives*, New York: Monthly Review Press.

Siembieda, W.J. and López Moreno, E. (1997) 'Expanding housing choices for the sector popular: strategies for Mexico', *Housing Policy Debate*, 8, 3: 651–78.

Sirivardana, S. (1986) 'Reflections on the implementation of the Million Houses Programme', *Habitat International*, 10, 3: 91–108.

—— (1994) 'Sri Lanka's experience with mobilisation strategy with special reference to housing and poverty alleviation', paper presented at the South Asian Regional Housing Conference, Colombo.

Smallman, P. (1977) 'The Housing Trust: a history of non-performance', *The Gleaner*, 3 December, Kingston.

Smets, P. (1996a) 'Community-based finance systems and their potential for urban self-help in a new South Africa', *Development Southern Africa*, 13, 2: 173–88

—— (1996b) 'Informal housing finance in Hyderabad, India', *Urban Research Working Paper* no. 40, Amsterdam: Vrije Universiteit.

——— (1997) 'Private housing finance in India: reaching down-market?', *Habitat International*, 21, 1: 1–15.

Smillie, I. (1995) *The Alms Bazaar: Altruism under Fire – Non-profit Organizations and International Development*, London: Intermediate Technology Publications.

Society for the Promotion of Area Resource Centres (SPARC) (1996) 'SPARC, the National Slum Dwellers Federation and Mahila Milan', in D. Mitlin (ed.) *Housing Finance and Resource Mobilization: Recent Innovations*, London: International Institute for Environment and Development.

Sparr, P. (1994) *Mortgaging Women's Lives: Feminist Critiques of Structural Adjustment*, London: Zed.

Srinivas, H. and Y. Higuchi (1996) 'A continuum of informality of credit: what can informal lenders teach us?', *Savings and Development*, 20, 2: 207–21.

Statistical Institute of Jamaica (STATIN) (1994) *Household Expenditure Survey, Volume II*, Kingston: STATIN.

Stevenson, R. (1979) 'Housing programs policies in Bogotá: an historical/descriptive analysis', World Bank City Study Research Project RPO. 671–47. Washington, DC: World Bank.

Strassman, W.P. (1982) *The Transformation of Urban Housing: The Experience of Upgrading in Cartagena*, Baltimore: Johns Hopkins University Press.

Struyk, R.J. and Turner, M.A. (1986) *Finance and Housing Quality in Two Developing Countries: Korea and the Philippines*, Washington, DC: Urban Institute Press.

Struyk, R.J., Katsura, H.M. and Mark, K. (1989) 'Who gets formal housing finance in Jordan?', *Review of Urban and Regional Development Studies*, 1: 23–36.

Subramanian, S. (1996) 'Vulnerability to price shocks under alternative policies in Cameroon', in D.E. Sahn (ed.) *Economic Reform and the Poor in Africa*, Oxford: Clarendon Press.

Tendler, J. (1989) 'Whatever happened to poverty alleviation?' *World Development*, 17, 7: 1033–44.

Thomas, E. (1991) 'Rotating credit associations in Cape Town', in E. Preston-Whyte and C.M. Rogerson (eds) *South Africa's Informal Economy*, Cape Town: Oxford University Press.

Tinker, I. (1992) 'Global policies regarding shelter for women: experiences of the UN Centre for Human Settlements', in H.C. Dandekar (ed.) *Shelter, Women and Development: First and Third World Perspectives*, Ann Arbor: George Wahr.

Titus, M. (1997) 'Developing financial services for the urban poor: the Sharan experience', *Environment and Urbanization*, 9, 1: 227–32.

Todes, A. and Walker, N. (1992) 'Women and housing policy in South Africa: a discussion of Durban case studies', in H.C. Dandekar (ed.) *Shelter, Women and Development: First and Third World Perspectives*, Ann Arbor: George Wahr.

Tomlinson, M.R. (1995) 'From principle to practice: implementers' views on the new housing subsidy scheme', *Centre for Policy Studies Research Report* no. 44. Johannesburg: Centre for Policy Studies.

Trialog (1995) *Special Issue on 'Community Based Housing Finance'*, 47, 4th quarter.

Turner, B. (ed.) (1988) *Building Community: A Third World Case Book*, London: Building Community Books.

Turner, J.F.C. (1967) 'Barriers and channels for housing development in modernizing countries', *Journal of the American Institute of Planners*, 33: 167–81.

——— (1968) 'Housing priorities, settlement patterns and urban development in modernizing countries', *Journal of the American Institute of Planners*, 34: 354–63.

—— (1976) *Housing by People: Towards Autonomy in Building Environments*, London: Marion Boyars.

—— (1982) 'Issues in self-help and self-managed housing', in P.M. Ward (ed.) *Self-help Housing: A Critique*. London: Mansell.

Turner, J.F.C. and Fichter, R. (1972) *Freedom to Build: Dweller Control of the Housing Process*, New York: Macmillan.

United Nations Centre for Human Settlementss (UNCHS) (1991a) *Housing-finance Manual for Developing Countries: A Methodology for Designing Housing-finance Institutions*, Training Materials Series, Nairobi: UNCHS.

—— (1991b) *Integrating Housing Finance into the National Finance Systems of Developing Countries: Exploring the Potentials and the Problems*, Nairobi: UNCHS.

—— (1993) *Public/Private Partnerships in Enabling Shelter Strategies*, Nairobi: UNCHS.

—— (1996) *An Urbanising World: Global Report on Human Settlements*, Oxford: Oxford University Press.

United Nations High Commission for Refugees (UNHCR) (1995) *In Search of Solutions: The State of the World's Refugees*, Oxford: Oxford University Press.

—— (1996) *Regional Conference to Address the Problems of Refugees, Displaced Persons, Other Forms of Involuntary Displacement and Returnees in the Countries of the Commonwealth of Independent States and Relevant Neighbouring States*, Geneva: UNHCR.

United Nations Development Programme (UNDP) (1992) *Human Development Report 1992*, Oxford: Oxford University Press.

United States Agency for International Development (USAID) (1997) *Building on Progress: The Future of Housing Finance in Poland*, Warsaw: USAID.

Urban Redevelopment Authority (URA) (1996) Real Estate Statistics Series: Stock and Occupancy, 3rd quarter.

Vakil, A.C. (1992) 'The contribution of community-based housing organisations to women's shelter and development: evidence from Zimbabwe', in H.C. Dandekar (ed.) *Shelter, Women and Development: First and Third World Perspectives*, Ann Arbor: George Wahr.

—— (1996) 'Understanding housing CBOs: comparative case studies from Zimbabwe', *Third World Planning Review*, 18, 3: 325–48.

Valença, M.M. (1992) 'The inevitable crisis of the Brazilian housing finance system', *Urban Studies*, 29, 1: 39–56.

van Huyck, A.P. (1987) 'The economies of shelter in development', in *Proceedings of the Second International Shelter Conference and Recommendations on Shelter and Urban Development*.

van Vliet, W. (1988) *Women, Housing and Community*, Aldershot: Avebury.

Vance, I. (1987) 'More than bricks and mortar: women's participation in self-help housing in Managua, Nicaragua', in C.O.N. Moser and L. Peake (eds) *Women, Human Settlements and Housing*, London: Tavistock.

Varley, A. (1987) 'The relationship between tenure legalization and housing improvements: evidence from Mexico City', *Development and Change*, 18, 3: 129–47.

—— (1994) 'Housing the household, holding the house', in G.A. Jones and P.M. Ward (eds) *Methodology for Land and Housing Market Analysis*, London: UCL Press.

—— (1995) 'Neither victims nor heroines: women, land and housing in Mexican cities', *Third World Planning Review*, 17, 2: 169–82.

—— (1996) 'Women heading households: some more equal than others?', *World Development*, 24, 3: 505–20.

von Pischke, J.D. (1996) 'Measuring the trade-off between outreach and sustainability of microenterprise lenders', *Journal of International Development*, 8, 2: 225–39.

Ward, P.M. (1981) 'Financing land acquisition for self-build housing schemes', *Third World Planning Review*, 3,1: 7–20.

—— (1991) 'Mexico', in W. van Vliet (ed.) *International Handbook of Housing Policies and Practices*, Westport: Greenwood Press.

Ward, P.M. and Macoloo, G.C. (1992) 'Articulation theory and self-help housing practice in the 1990s', *International Journal of Urban and Regional Research*, 16, 1: 60–80.

Ward, P.M., Jiménez, E. and Jones, G.A. (1993) 'Residential land price changes in Mexican cities and the affordability of land for low-income groups', *Urban Studies*, 30, 9: 1521–42.

Watson, S. (1988) *Accommodating Inequality: Gender and Housing*, Sydney: Allen & Unwin.

Wegelin, E.A. and Chanond, C. (1983) 'Home improvement, housing finance and security of tenure in Bangkok slums', in S. Angel, R. Archer, S. Tanphiphat and E. Wegelin (eds) *Land for Housing the Poor*, Singapore: Select Books.

Wilkie, J., Ochoa, E. and Lorey, D.E. (1990) *Statistical Abstract for Latin America 28*, Los Angeles: University of California Press.

Witter, M. (1996) 'Between macroeconomic stability and microeconomic instability: the housing sector in Jamaica', in A. Francis, V. George and M. Smith (eds) *Affordable Housing: The Continuing Search for Solutions*, Proceedings of the National Housing Trust's 20th Anniversary Housing Symposium, Kingston: NHT.

Women's Construction Collective (WCC) (1989) Video documentary produced and directed by the Inter-American Foundation, Rosslyn, Virginia.

Wong, A.K. and Yeh, S.H.K. (1985) *Housing a Nation: 25 Years of Public Housing in Singapore*, Singapore: Maruzen.

World Bank (1975) *Housing: Sector Policy Paper*, Washington, DC: World Bank.

—— (1993a) *Housing: Enabling Markets to Work, World Bank Policy Paper*, Washington, DC: World Bank.

—— (1993b) *The Housing Indicators Program: Preliminary Results*, Washington, DC: World Bank and UNCHS.

—— (1995) *World Development Report 1995: Workers in an Integrating World*, Oxford: Oxford University Press.

—— (1997) *World Development Report 1997: The State in a Changing World*, Oxford: Oxford University Press.

Yaron, J. (1992) 'Successful rural finance institutions', *World Bank Discussion Paper* no. 150. Washington, DC: World Bank.

Yen, M.M. and Keigher, S.M. (1992) 'Shelter options for elderly women: an overview', in H.C. Dandekar (ed.) *Shelter, Women and Development: First and Third World Perspectives*, Ann Arbor: George Wahr.

Young, K. (1993) *Planning Development with Women: Making a World of Difference*, New York: St Martin's Press.

Zamora Rivas, M. (1993) *De la guerra a los acuerdos de paz: la mujer en el área urbana de El Salvador, Del trabajo no-remunerado al trabajo productivo/La participación de la mujer en el sector informal urbano*, San José, Costa Rica: Fundación Arias para la Paz y el Progreso Humano.

Zaobao, L. (1988) *Survey of Youths on Basic Values*, Singapore: Singapore Press Holdings.

Zayonchkovskaya, Z., Kocharyan, A. and Vitkovskaya, G. (1993) 'Forced migration and ethnic processes in the former Soviet Union', in R. Black and V. Robinson (eds) *Geography and Refugees: Patterns and Processes of Change*, London and New York: Belhaven.

Zearley, T.L. (1993) 'Creating an enabling environment for housing: recent reforms in Mexico', *Housing Policy Debate*, 4, 2: 239–48.

Zepeda Payeras, M. (1996) 'Considerations on the strategy of a secondary mortgage market in Mexico', paper presented at Mexico/US Global Forum, Aspen, Colorado.

Ziccardi, A. and Mier y Terán, A. (1990) 'La producción de vivienda del sector público en México 1980–1990', *Vivienda*, 1, 1–2: 6–13.

INDEX